T0320993

Disability Rehabilitation
Management through ICT

RIVER PUBLISHERS SERIES IN COMMUNICATIONS

Consulting Series Editors

MARINA RUGGIERI
University of Roma "Tor Vergata"
Italy

HOMAYOUN NIKOOKAR
Delft University of Technology
The Netherlands

This series focuses on communications science and technology. This includes the theory and use of systems involving all terminals, computers, and information processors; wired and wireless networks; and network layouts, procontentsols, architectures, and implementations.

Furthermore, developments toward new market demands in systems, products, and technologies such as personal communications services, multimedia systems, enterprise networks, and optical communications systems.

- Wireless Communications
- Networks
- Security
- Antennas & Propagation
- Microwaves
- Software Defined Radio

For a list of other books in this series, please visit www.riverpublishers.com

Disability Rehabilitation Management through ICT

Editors

Dr. M. D. Tiwari

Director IIIT,
Allahabad & RGIIT, Amethi
and
Former President
Association of Indian Universities
New Delhi

Dr. Seema Shah

Deputy Registrar
IIIT, Allahabad
U.P, India

Dr. Iti Tiwari

Associate Professor
UPRTOU, Allahabad

River Publishers

Aalborg

Published, sold and distributed by:
River Publishers
PO box 1657
Algade 42
9000 Aalborg
Denmark
Tel.: +4536953197

www.riverpublishers.com

ISBN: 978-87-92329-49-3

Preface

Emerging technologies have been successfully employed within different disciplines improving the quality of life globally In India these have been employed to address the multifarious issues in a limited manner to improve the way of life for individuals with different disabilities. Indian Institute of Information Technology, Allahabad, therefore, organized International and National Seminars since 2008 to deliberate on the application of information communication technologies [ICT] in the rehabilitation of people with disabilities

Domain experts have recognized that management of differently abled individuals requires participation of multidisciplinary experts. We also require an enormous bank of skilled, qualified trained manpower with support to keep themselves in tune with latest technology and developments. We also face the problems of having to deal with individuals located countrywide; inadequate ICT-based infrastructure to assist with accessing services of experts that are based in urban areas. This is one of the major constraints for a sustained and effective therapeutic program.

The major issues inhibiting an effective sustained and broad based program identified in the Seminars are:

- Ongoing efforts are required to usher in positive change within the care provider; community; members of the family and government support.
- We have a limited and inadequate number of skilled personnel, experts and institutions to take care of the growing population.
- Extremely poor accessibility and connectivity to reach out the services and guidance of experts and institutions

The articles cover the following four areas on disability rehabilitation management:

- Rehabilitation of Differently Abled Persons: Overview and Issues
- Infrastructure and potentials for use of ICT in Disability Rehabilitation
- ICT Initiatives for Rehabilitation of Differently Abled Persons
- Assistive Technology for Differently Abled Persons — the road ahead

This is the fifth publication under the IIIT-A Series on e-Governance. It is a collection of 20 articles based on the presentations made in the Seminars. This book will of interest to all stakeholders in the disability rehabilitation management as the population of people with disabilities in growing.

We would like to convey our thanks and appreciation to all eminent contributors for making available their valuable time to participate in the deliberations of the Seminars and their contribution for this book. We are confident that this is the beginning of multi-disciplinary domain experts come on one platform to bring people with disabilities as productive members of our society.

M D Tiwari
Seema Shah
Iti Tiwari

Editors' Biography

Dr. Murli Dhar Tiwari was born on August 16th, 1948 in a small village near Amethi in an ordinary family. After school education he completed his B.Sc., M.Sc. and D.Phil. Degrees from Allahabad University. In 1973 he joined HNB Garhwal University, India as a Lecturer, subsequently he became Head of Physics Department of the same University. In 1984 he joined University Grants Commission, New Delhi as a Principal Scientific Officer and after working for a decade on senior position there he moved to All India Council for Technical Education, New Delhi as a Senior Advisor in 1994.

In 1995 he joined MJP Rohilkhand University, Bareilly, India as a Vice Chancellor. During 3 years of his tenure he managed a number of Professional Courses at this University which was adjudged First University in all Universities in performance appraisal in three consecutive years. He was also honoured by substantial financial support from the Government. He is also Founder Chairman of U.P.State Council of Higher Education. He has enjoyed prestigious fellowships like **Alexander-von-humboldt Fellowship** and visited abroad several times for academic endeavours.

Dr. Tiwari moved to Allahabad, India and established Indian Institute of Information Technology, Allahabad (IIIT-A) in 1999. Since then he is working there as a Director.

The Institute got collaborations with most of the top class Universities of USA, Switzerland, U.K. and others. He has published about 150 research papers in refereed journals, supervised 12 Ph.D. students and has written/edited about 12 books published by publishers of International repute.

Dr. Tiwari has also worked on prestigious positions such as President of Association of Indian Universities, New Delhi. At present he is Chairman, Electronics and Technology Division, Bureau of Indian Standards, New Delhi,

Chairman of several bilateral science and technology programmes of Government of India.

To promote science education and research in the country, Dr. Tiwari is organising Conclave of Nobel Laureates since 2008 every year and the same is being done in 2012. This is a unique programme and is contributing a lot for awareness and interests of young brilliant students for science education & research.

At present his interests are Information Technology in general — Wireless Sensors, Human Computer Interaction and Application of IT in untaped Non-conventional Energy in particular.

Dr. Seema Shah has been awarded D.Phil degree from the famous university of Allahabad, U.P., India in the area of Human Resource Management. She is crowned with gold medal of being topper of P.G. in Human Resource Management of M.P. Bhoj Open University, Bhopal India. She has already acquired Master Degree from Kumaun University, Nainital. She has also been honoured as a best employee in the city of Allahabad. Her knowledge of Application of ICT in Human Resource Management and other related matters have been par-excellence elsewhere by developing new softwares. She has already published several research papers in International journal of world repute, which have been copyrighted by Microsoft and Google Scholar as well. They have been cited by several authors globally. She has also published an edited book and organized several National/International Conferences/Seminars. Her interest is in the Application of ICT in HRM of educational institutions. She is working in Indian Institute of Information Technology, Allahabad, U.P., India as a Deputy Registrar.

Dr. Iti Tiwari is Associate Professor in Uttar Pradesh Rajarshi Tandon Open University, Allahabad, India. Prior to that she has Reader & Head, Sociology Department at Jagadguru Rambhadracharya Handicapped University, Chitrakoot, Faculty in Rakshapal Bahadur Institute of Management, Bareilly and Coordinator (Distance Education), G.G.D. University, Bilaspur. Her specialization is in Child Crime and Handicapped. She has tremendous contributions in the area of physically challenged persons and Child crime. She has published more than 150 articles and manuscripts, 3 books and substantial research publications.

Contents

Part I

Rehabilitation of Differently Abled Persons: Overview and Issues

1

Raising the Floor: Providing Accessible Technology and Content to the Disabled*

James R. Fruchterman

The goal of Raising the Floor Initiative is to ensure that every person with a disability on the planet has access to the basic technologies and knowledge they need to fully participate in education, employment and social inclusion. The cell phone revolution combined with publicly available computers and servers on the Internet have created an opportunity to meet the needs of the poorest of the poor in the world, for free or a locally affordable price. And it's imperative that we do it.

Raising the Floor is not about charity, although it will require nonprofit action. It's not about making money, although active participation of business is the only way to reach the entire planet. Government can encourage equal opportunity, but can't deliver it alone. Raising the Floor is all about providing access to the tools that people with disabilities need to independently advance in society. Business alone hasn't done it. Charity alone hasn't done it. Government alone can't do it. Blending the power of these sectors through social entrepreneurship can bridge the gaps and make this happen in the next decade.

---

*This paper is based on a keynote presentation given at the International Conference on Computers Helping People with Special Needs, Linz, Austria, July 2008.

This keynote presentation will provide background on how Benetech was founded as a social enterprise for using technology to serve humanity, and how we've succeeded in bridging the gap between highly profitable and socially important technology applications.

Rocket Science and Reading Machines

My personal plan was to become an astronaut. While an undergraduate at the California Institute of Technology, I took classes in both aerospace and electrical engineering, as well as applied physics and computer science. My 'light bulb moment' came in a modern optics class where we were discussing developing pattern recognition systems. Since this was the 1970s, the example was drawn from the military. We learned how you could build a smart missile with a camera in the nose, which used optical pattern recognition to home in on a target such as a tank and blow it up. I left the class wondering if there might be a more socially beneficial application of this technology!

My idea was to use the same kind of technology to build a reading machine for the blind. It would do optical character recognition with a spinning disk with a special pattern for each letter, and a phototransistor would record the signal when the letter being analysed correlated with a specific pattern on the film disk. This was a long time ago.

My professor, Bert Hesselink (now at Stanford), was encouraging but pointed out some of the practical limitations of this implementation. Still, it was cherished as the one good idea I had in college. I followed Bert Hesselink to Stanford to start my Ph.D., but quickly became sidetracked by the lure of the Silicon Valley.

The company that enticed me from my doctoral studies was a private rocket company. This was an irresistible opportunity given my space inclinations, and I took a leave from Stanford to help build a rocket. Within six months, the rocket had blown up on the launch pad and I was looking for something new to do.

My boss from the rocket company, David Ross, introduced me to one of his friends who was a chip designer from HP Labs. We sat down in 1981 to talk about starting a company. The chip designer, Eric Hannah, described his idea of making a custom chip that could do optical character recognition. I was enchanted: my first reaction was: 'You could make a reading machine for the blind with that'!

Calera Recognition Systems

Eric Hannah, David Ross and I founded the company that became Calera Recognition Systems in 1982. We raised more than U.S. $25 million through venture capital during the 1980s, which was a lot of money then (and now). We promised our investors that we would make a machine that would read anything, and then set forth to do it. We were lucky, because we happened to start the company just at the time when it became computationally possible to make such a machine.

At the time, optical character recognition was a field populated by more than fifty companies. The leading technologies were font-specific. You bought a machine from DEST that came with PROMs (Programmable Read-Only Memory) that each recorded the pattern of a specific type font. You could buy a PROM for Courier, or Prestige Elite, or one of the OCR-specific fonts that had been purpose-built for this task. However, this approach was no longer working. The world of typography was changing with the advent of desktop publishing and laser printers, where people were using proportional spacing (rather than the old fixed-width typewriter fonts) and where companies were inventing new fonts all the time. A different kind of OCR was needed.

Our closest competitor was Xerox, which had purchased Ray Kurzweil's company a few years before we got started. Ray had created a machine that could read anything, but to get good performance you needed to train the machine on the font on the document you wished to read. Our challenge was to make omni-font character recognition really work without needing human training.

Calera went on to succeed commercially in character recognition, and is now part of Nuance, the leading company in the retail OCR field. But we were interested in the trendiest application of the technology, which was helping people with disabilities.

A machine that could read just about any font was essential for creating a usable reading machine for the blind. Calera's engineers and marketing people were very excited about this project, and built a prototype reading machine for the blind as a side project in 1987. The prototype would scan a page, do the optical character recognition, deliver the recognized text to a PC, and then a (very) mechanical synthetic voice would read the words aloud.

The crucial moment came when we demonstrated this prototype to our venture investors. They were impressed by the capabilities and asked about the size of the market for reading machines for the blind. As the vice-president of marketing, my answer was that it was roughly $1 million dollars per year. Our investors were unimpressed with this: they had invested more than $25 million because we had promised them a $100+ million market. Good public and employee relations or social responsibility was not enough: the investors did not want us to spend the time to launch this product. They understandably wanted us to focus on the money-making opportunities we had initially promised them.

Arkenstone

Dropping the reading machine because of lack of investor enthusiasm was not an acceptable option to me. We soon started a deliberately nonprofit technology company as opposed to the accidentally nonprofit technology companies of the Silicon Valley. We created a charity named Arkenstone, and then created a business inside it of selling reading machines for the blind. We assumed we could build a breakeven nonprofit organization with revenues of $1 million per year.

Within three years of its founding, Arkenstone was profitably generating $5 million a year in revenues. By dropping the price of a reading system to less than $5,000, many more blind people could afford them. The key design choice was to make the reading machine on top of a standard personal computer. During the 1990s, this enabled the price of a reading system to go from $5,000 to $4,000 to $3,000 to $2,000 to $1,200. Each time the price fell, more people could afford to get reading machines. By 2000, Arkenstone had provided over 35,000 reading systems, reading a dozen languages, in sixty countries around the world.

The main engine of these price decreases was the personal computer industry's habit of relentlessly delivering more for less. When we started in 1989, the scanner, OCR co-processor card and voice synthesizer were all hardware devices that cost over $1,000 each. By the end of the 1990s, the scanner cost less than $100 and the OCR and voice synthesizer were software packages that could be obtained for much less money than before.

Interestingly, while Arkenstone was designed solely as a reading machine for the blind, when we surveyed our early customers we discovered that nearly 20 per cent of our users were not blind but dyslexic. The original software highlighted each word as it was spoken providing a multi-modal model that was ideal for dyslexic users. We were able to quickly develop a new product with a user interface specifically designed for these users with learning disabilities. We were then better able to serve this sector of the disability market.

Benetech

Seeing the value in providing affordable tools that greatly empowered the reading disabled, I was inspired to try to replicate our success with multiple enterprises that utilized technology to serve social causes. In early 2000, Arkenstone's business assets were sold to a for-profit company, Freedom Scientific, and we had to rename our non-profit Benetech. The resulting money went into creating new-technology social enterprises. Our goal was simple: to create new technology solutions which serve humanity and empower people to improve their lives.

One of our first focus areas was human rights. We recognized that human rights activists have only one asset — information. A documented record of human rights violations is the only effective tool there is against the perpetrators.

The information can be used to 'name and shame' putting pressure on those allied with the violators as well as creating a record that will, hopefully, lead to a day in court where the perpetrators are convicted of crimes against humanity.

As you would expect, human rights groups throughout the world gather massive amounts of violation data. Unfortunately, much of it never reaches its full potential or intended audience. Instead, a large part of this information is lost due to confiscation or destruction, neglect, passage of time, or because the grassroots organizations that collect the data lack the resources or infrastructure to document and communicate violations systematically and securely.

In 2003, Benetech launched Martus, an information and document management system based on client software and Internet-based infrastructure. These tools help to ensure that the documentation of human rights violations is safeguarded and disseminated. This, in turn, strengthens reporting of violations, and in some cases, has prevented additional abuses. Martus is a tool designed for and used by the activists 'on the ground' gathering testimony and stories about human suffering.

Such formal collection and collation of data and large-scale analysis can prove that many cases of mass violence are not isolated incidents, but rather planned and systematically executed policy. Such findings can build strong, defensible claims about what has been endured by victims and societies. The result: perpetrators can be brought to justice, which hopefully will prevent the recurrence of atrocities.

Benetech's Human Rights Data Analysis Group was involved with the analysis and prosecution of war crimes in the former Yugoslavia. Since we had done an analysis of all the people who had left Kosovo, it was possible to actually reconstruct when they left. We tested the defense theory that it was NATO bombing that caused them to leave their villages and the prosecution's theory which was killings by forces allied with Serbia. The graphic of people leaving closely synchronized with the graph of people being killed and not with NATO bombing.

Our human rights work is an example of technology serving humanity. If we can make activists more powerful and help them pursue human rights and social justice, then we've been successful in reaching our goals.

Bookshare.org

Our first online disability project under the new Benetech umbrella was inspired when my 14-year-old son introduced me to Napster. It occurred to me that this type of peer-to-peer model could be used to help with the accessible book challenge.

Having developed the Arkenstone reader, we understood that the last thing that a print-disabled person wanted to do when she wanted to read a book was to spend three hours scanning it first. We conceived of the concept of a digital library built by the people who use the library, instead of by librarians deciding what disabled people should read. The result was Bookshare.org — named to emphasize the social values of how the library would be built — and it was the next generation of solutions for the disabled community.

A simple description of Bookshare.org is a website that is a cross of Amazon.com, Napster and Talking Books for the blind, but legal. Volunteers scan books using both standard commercial OCR packages (like OmniPage and

FineReader) as well as specialized reading software for the disabled (like Kurzweil, WYNN and Open Book). Using a provision of the US copyright law, they upload these scanned books to a website called Bookshare.org, where people with bona fide print disabilities can download the books. Print disabilities are those that make picking up and reading a printed book difficult, such as blindness, low vision, severe dyslexia and certain physical disabilities such as quadriplegia or cerebral palsy. Once downloaded, these digital books can be turned into an accessible form. Typically, this is synthesized speech output, but it can also be enlarged for those with low vision as well as output in Braille.

Thanks to its community of on-line volunteers, Bookshare.org now has over 95,000 books available to print-disabled Americans. The majority of the books in the library were scanned by a person with a disability and, in all likelihood, a person with a disability proof-read it and added it to the collection.

Traditionally, people with disabilities don't get to choose which books are made available in accessible form. Because of the expense to provide books using traditional approaches, there are only a small percentage of books ever prepared in an accessible format. The Bookshare.org approach is open ended — if someone in the community feels it is worth their time to scan that book, we'll add it to our collection. And because we are doing the first completely virtual library, it costs us almost nothing to add a book to our servers.

Our plan was to charge US $50/year for an unlimited library card to Bookshare.org and within a couple of years that would be enough to pay our costs. After five years, we had funded only one-third of our budget through user subscriptions. The rest was coming from individuals and foundations. Due to the copyright law, we were essentially restricted to serving only the print-disabled community in the United States. Due to the cost, we knew we were only serving a small portion of the US disability market. However, we are working to change this and make Bookshare.org much more widely available both inside and outside the United States.

Bookshare for Education

In 2004, legislators passed a law in the US mandating that all K-12 students with print disabilities have access to textbooks in an accessible form for free. The law effectively made it a requirement for publishers to create a high-quality version of textbooks in a national standard format based on the DAISY standard — the same format that is already being delivered by Bookshare.org. Three years after this law was passed, almost no student had gotten any books through this new law. It was very frustrating for all involved. In 2007, the US Federal Department of Education held a competition to essentially do what the law requires, which is to deliver accessible books to all students for free including specialized assistive technology. Benetech's proposal, Bookshare for Education, was chosen and we were awarded a $32 million contract to deliver these services over the next five years.

A key piece of Bookshare for Education is partnerships with the assistive technology industry. Benetech has had a long-term partnership with HumanWare to provide Bookshare.org users with a version of the Victor Reader software for no additional cost. We'll be working with HumanWare to make new improvements, which will be available to Bookshare.org users globally.

A new partnership with Don Johnston for a special version of their Read:OutLoud software will provide every user with a print disability that qualifies for Bookshare either a free copy of Read:OutLoud or Victor Reader software. Again, is it free? No, the US government is paying for it, or subscribers are paying for it. But, that's not enough. If our goal is to address all disabled people in the world, we need to provide solutions for the person who doesn't have a PC at home. These users should be able to use free public access terminals in the library that are not equipped with any assistive technology but are capable of accessing Bookshare.org through a web-based interface. In the longer term, the goal will be providing access through MP3 players or cell phones which are widely used by people with disabilities. Imagine books available as MP3 files and accessible through the low-cost MP3 players and cell phones that users are already using today.

Lastly, we are now actively engaging publishers in the question of permissions. To serve people with disabilities outside the United States, we can no longer depend on a domestic copyright exemption. Although advocates are working on changing international norms, and coming up with copyright policies that encourage cross-border sharing of accessible materials, this is not going to happen soon. Many publishers are comfortable with the idea of supporting access for people with disabilities, and we have already been successful in gathering global rights from a number of leading American publishers, such as Scholastic and HarperCollins. In addition, we've built partnerships with several publishers in India who are already granting us global permissions.

To make Bookshare.org fully accessible globally, we need to do three important tasks. We need to gain access to the content legally, through our permissions drive. We need to make Bookshare.org more affordable to consumers, especially those in the less developed countries. Lastly, we need to make accessible technology much more practical for the typical global citizen with a disability. We see this all as part of 'Raising the Floor'. It's much broader than just book access.

Raising the Floor

The Raising the Floor concept came out of conversations with Dr. Gregg Vanderheiden from the Trace R&D Center at the University of Wisconsin, Madison. The essential core of Raising the Floor is ensuring every disabled person on this planet has access to assistive technology and the content he/she needs to pursue education, employment and social inclusion.

Our goal is to create the same kind of dynamic that exists in the market for items such as PCs and cell phones. In mass markets, when prices decrease by a factor of ten, sales go up by a factor of 100, or 1,000, or even 10,000. Price declines like this are very difficult for the assistive technology industry to achieve. In the past we've been proud of selling $1,000 products, because at the time that was a dramatic price decrease. But the reality is assistive technology is one of the few technology sectors where prices have effectively gone up over the last decade.

If one actually stands back and looks at the entire world, the assistive technology industry today cannot be serving more than 1 per cent or 2 per cent of people with disabilities. The industry has great products that do great things but which are beyond the reach of the majority of people with disabilities. Even in the United States, there are a vast number of families with a disabled child that do not even own a personal computer. If you consider the developing world, those numbers really make you think.

Raising the Floor is not about making high-priced technology free. It's about providing a basic level of functionality to all, while allowing those who are economically empowered the option to enhance their solutions. In many societies, we provide free public libraries because we feel that information should be freely available to all. We provide free public education because we feel that education should be available to all. In a similar manner, we must commit to providing people with disabilities at least the basic assistive technologies needed to enable equal access to the vast amount of information, resources and services available online to everyone else for life's essential activities.

Poor people need access to clean water and they can't afford to pay $2.50 for a plastic bottle of water, but when people become economically empowered there is a large market for $2.50 bottles of water. There is room for a field where core assistive technology is like clean tap water: abundant, basic, affordable and life-giving. With access to education and employment, the market for premium assistive technologies will expand as more people with disabilities will be empowered economically and will demand more.

We as technologists working in the field need to recognize that people with disabilities do not exist to serve our need to make money or to be an organization operating in perpetuity. We exist to serve people first, and Raising the Floor is about galvanizing this transition where assistive technology fundamentally shifts to be far more available. This shift will be based on three trends we can already see in the mainstream technology markets.

The biggest trend is already here — cell phones. Cell phones will be the default platform. There are three billion cell phones in use right now. Half the people on the planet have a cell phone. It's the most successful piece of technology to reach the poor; the most successful piece of assistive technology ever invented. The cellular phone may even be more popular with blind people than the white cane.

The network is the second piece. Cell phones allow us access to the network. Through the network, we can connect to much of the world's information, to the many capabilities of the computer servers on the Internet and, most important of all, to the majority of humanity. By connecting people with disabilities, we raise their collective capabilities. Why is Bookshare.org the fastest growing library for people with disabilities? Because we can harness the interests and efforts of thousands of people at almost no cost. This kind of scale is ready to be harnessed.

The last and third piece is the concept of open source and open content. Firefox is a great example of an open source program that has a social impact. The business model is that Firefox is free. It was built by people who wanted to offer an alternative to Microsoft's Internet Explorer. They did it out of passion for service. Most people love to serve other people. The open source movement provides a legal framework that allows this passion for service to be dedicated to serving the public good, rather than private aims.

There is also a similar growing movement in open content. Wikipedia is a great example and there will be many more in the future. Creative Commons is a nonprofit founded solely to foster the sharing of creative works. It provides a structure allowing anyone to create something and, rather then restricting access with copyrights, encourages sharing it under generous terms.

Conclusion

Raising the Floor is a vision of the future where the majority of people with disabilities have access to the tools and content they need to actually go forth to get education and realize economic opportunity — to fully participate in society. This is not anything more than permitting the potential of people with disabilities to fully blossom.

It is not telling disabled people to adapt to existing assistive technology but to actually ask them, 'How can we make this technology relevant in your life and get you what you need based on where you are and what you have?' Our job as technologists is to knock the barriers out of the way and let the users of assistive technology make what they will of their lives. Users will tell us what we need to do to make that possible.

The assistive technology sector will need to come along. We're going to find a way to define new business models such as those that have already happened to mainstream technology. Companies are already giving away free software and services in mainstream technology and finding a way to make money, such as Google. We can do that as a sector and we'll do it while making available viable careers for people in the industry and ensuring the social justice objective of people having the access that equity and equality demand.

Raising the Floor must be a global initiative. It will take much more than one company or organization. If we truly believe in social justice and inclusion, we will work together to ensure that the fruits of technology reach every person with a disability in the world.

2

Assistive Technology for the Differently-Abled

B Woodward

Emerging communication technologies are now being exploited not just for speech transmission but also for data transmission. Mobile phones, with their multi-function capability, although designed principally for speech communications, can now send digital photographs, e-mails and even data representing medical signals. In India, data can similarly be transmitted by a freely-available satellite service or by a fibre-optic network. This assistive technology, focused on monitoring people's well-being remotely, is now available to improve the quality of the lives of people with disabilities. However, the way this might be achieved is far from clear, especially given India's size and population, with many people living in areas remote from hospitals and care centres. Nevertheless, some interesting possibilities have been suggested in this essay.

Introduction

The usual purpose of telemedicine, or e-health, is to transmit biomedical data from a patient to a doctor without the need for both people to be in the same physical location [1]. Typical data can be an electrocardiogram (ECG), blood pressure, oxygen saturation or blood glucose level [2]; other data might be in the form of images, such as X-rays or ultra-sonograms [3]. The data may be viewed either for long-term monitoring of a chronic disease or for post-operative checks. According to conventional wisdom in developed countries, telemedicine represents a considerable saving in costs since patients can be monitored remotely rather than having to attend a hospital. Telemedicine is therefore beneficial to taxpayers who fund public health services. Another rationale for promoting telemedicine is that it saves significant time, particularly the patient's time, because it obviates the need to travel to see a doctor, often for a brief appointment. Governments are now tending to encourage a patient-centred approach to healthcare, by devolving primary care away from doctors, so that nurses or paramedics in the community obtain diagnostic data that are transmitted to a hospital [4].

However, the situation differs in India, with the number of patients being counted in tens or even hundreds of millions. Further, the distances separating these patients from a medical centre may preclude them from getting there. For patients with disabilities, who are perhaps also suffering from long-term illness, their needs are difficult to assess, let alone satisfy. It is therefore worthwhile to consider which technologies are available and how in due course these might be used to address the problem for the significant minority of people who are differently-abled. One approach might be that each person to be monitored carries a 'body area network' (BAN), comprising one or more 'intelligent' sensors connected to a processor module that may be worn as 'smart clothing'. Data from the BAN are then transmitted to a health professional (doctor, nurse or paramedic) who assesses that person's health or physical status.

For an out-patient who is not bed-bound but free to move around normally, communications technology enables the data to be transmitted using a mobile phone [5–7]. Wireless telemedicine is therefore inherently flexible in application, although a mobile phone itself would be beyond the means of most of India's rural poor population. In terms of the technology, a variety of configurations is possible, and some of these may eventually be made extremely small using microelectronics techniques. The next step requires careful thought on the part of technologists and clinicians, working together to decide and possibly prioritise the applications and modes of operation. How data is transmitted does not matter from a patient's point of view, although it is implicit here that wireless telemedicine will be adopted in at least part of the transmission link, since a huge number of patients, particularly those with disabilities who live in villages, could benefit from regular remote monitoring.

While the transmission of biomedical data by satellite is demonstrably feasible for long-range wireless links [8–10], it is more likely that mobile phones using cellular networks will be adopted for transmission over shorter ranges. Among related technical problems is the issue of how to link the data from the patient to the phone. This may possibly be done by a direct connection, or by a wireless connection, such as an acoustic, infra-red or *Bluetooth* radio link [5–7,11]. The potential advantage of *Bluetooth* and other wireless technologies is well recognised, including the Wireless Application Protocol (WAP) for cell phones, and Personal Digital Assistants (PDA) [11].

What exactly is needed to assist and empower differently-abled people, given the availability of communications technologies and now-ubiquitous computersc? Monitoring their health status is certainly important, notably for patients with long-term illnesses such as heart disease, diabetes or asthma, or with medical impairments such as neurological, visual, hearing- or speech-related, or with mobility problems through injury or disease. So using telemedicine as a means of monitoring these conditions may be seen as an important advance in rehabilitation or recovery.

Methodology

The telemedicine system developed at Loughborough University is capable of transmitting up to four sets of biomedical data from a patient via the General Packet Radio Service (GPRS) to a medical centre [5–7]. The system, shown in Figure 2.1, comprises a mobile patient unit, which is connected to the rest of the system via a GPRS Base Station, then over conventional land lines, i.e., the Public Switched Telephone Network (PSTN), to an Internet Service Provider and finally to computers on a local area network (LAN) in a hospital.

The patient mobile unit comprises an embedded processor that acquires real-time data from biomedical sensors. The data are stored temporarily in a 10-minute rolling memory that is over-written continuously. Interfacing is necessary to provide a Bluetooth interface with the phone, and to provide the patient with status messages.

The data are transferred from the memory of the mobile patient unit, via the GPRS network, to a server in the hospital, where they are archived and made available to a clinician as and when required. In an auto-answer mode of operation, the clinician can dial the patient's telephone number and download the

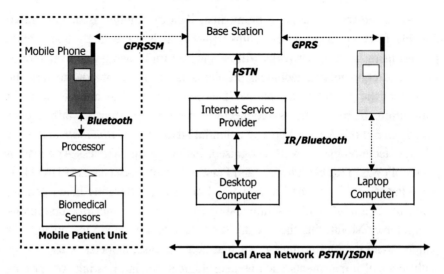

Fig. 2.1 Telemedicine system with mobile telephone and PSTN communications links.

data automatically. Acknowledgment messages are then sent from the server to the mobile unit to indicate successful receipt of a packet or a demand for re-transmission. The patient data include the International Mobile Subscriber Identity (IMSI) number, which is unique to the Subscriber Identity Module (SIM) card in the mobile telephone. This number is used to identify the patient to the system and to permit the storing of additional records for that patient. Once identified to the database, the clinician can display the patient's data and medical history.

The system is designed to be generic, so it may also be used for non-medical applications, including various 'assistive technologies' considered here, and to be 'future-proofed' in view of constantly evolving telecommunications technology, including the introduction of the Third Generation (3G) and higher protocols.

Discussion

Developments in high-speed wireless communications are likely to become more important in future telemedicine applications. At present, the Global System for Mobile Communications (GSM) is the main network in India. The alternative is GPRS, which is a digital system whose architecture is based on the Integrated Services Data Network (ISDN). Whereas ISDN operates with a basic rate of 64 kbit/s, GSM has a full channel rate of only 22.8 kbit/s, which includes significant coding. The coding has proved to be a useful feature in the work described here because although the actual data rate is limited to 9.6 kbit/s, the probability of error transmissions is greatly reduced. It is expected that GSM will eventually be replaced by GPRS and possibly by 3G, which will have a much greater data rate capability (144 kbit/s), comparable with that for the *Bluetooth* technology (434 kbit/s), which may be used for short-range telemedicine applications. Although these developments are unlikely for many years on a country-wide scale in India, they would allow much more data to be transmitted and would be particularly suited to the transmission of several channels of bio-medical and physical data.

Conclusion

The main contribution of this paper has been to describe the present scope and future potential of mobile communications in telemedicine. A modular-structured system has been used to illustrate the concept. The mobile patient unit comprises a processing unit, which accepts signals from one or more sensors, linked by a wireless channel to a mobile telephone. The prototype version is designed to transmit digitised electrocardiogram signals to a hospital via the GPRS mobile telephone cellular network. The modular design of the system should enable telemedicine providers to adapt to future technology advances. Any medical or physical parameter can be transmitted based on this system. This makes it suitable for assisting differently-abled people in countries with very large populations, such as India. There is a serious downside to the advent of this technology, which should not be overlooked: anyone who is provided with instrumentation that is effectively 'all-seeing' may be able to exercise surveillance, which could possibly be intrusive and may make people very uncomfortable. Thus, an element of sensitivity is essential when designing any system. For instance, a patient-operated OFF switch would have to be a fundamental feature of the design of any system used by this author!

References

[1] Wootton, R and J. Craig, *Introduction to Telemedicine*, Royal Society of Medicine Press Ltd, 1999.

[2] Freedman, S.B., 'Direct transmission of electrocardiograms to a mobile phone for management of a patient with acute myocardial infarction', *J. Telemed. Telecare*, Vol. 5, 1999, pp. 67–69.

[3] Laminen, h., L. Salminen, J. Makela, M. Lampinen and H. Uusitalo, 'Wireless picture transfer as a tool of primary health care', *J. Telemed. Telecare*, Vol. 5, no. 4, 1999, pp. 260–261.

[4] Kendall, L., 'The Future Patient', Institute of Public Policy Research, 2001; http://www.bl.uk/welfarereform/issue25/health.html

[5] Woodward, B., R. S. H. Istepanian, and C. I. Richards, 'Design of a telemedicine system using a mobile telephone', *IEEE Trans. Info. Technol. in Biomed.*, Vol. 5, March 2001, pp. 13–15.

[6] Rasid, M.F.A and B. Woodward, 'Bluetooth telemedicine processor for multichannel biomedical signal transmission via mobile cellular networks', *IEEE Transactions on Information Technology in Biomedicine* (Special Issue on e-Health), vol. 9, 2005, pp. 35–43.

[7] Woodward, B., M. F. A. Rasid, L. Gore and P. Atkins, 'GPRS-based mobile telemedicine system', *Journal of Mobile Multimedia* (Special Issue on Advanced Mobile Technologies for Health Care Applications), Vol. 2, no. 1, 2006, pp. 2–22.

[8] Satava, R., P. B. Angood, B. Harnett, C. Macedonia and R. Merrell, 'The physiologic cipher at altitude: telemedicine and real-time monitoring of climbers on Mount Everest', *Telemed. J. and eHealth*, Vol. 6, no. 3, 2000, pp. 303–313.

[9] Boyd, S.Y.N., J. R. Bulgrin, R. Woods, T. Morris, B. J. Rubal and T. D. Bauch, 'Remote echocardiography via INMARSAT satellite telephone', *J. Telemed. Telecare*, Vol. 6, 2000, pp. 305–307.

[10] Laminen, H., 'Mobile satellite systems', *J. Telemed. Telecare*, Vol. 5, no. 2, 1999, pp. 71–83.

[11] Bates, J., B. Demuth, V. Pendleton and K. P. Tonkin, 'Wireless applications in mobile telemedicine', Abstracts from The American Telemedicine Association 6th Annual Meeting, Fort Lauderdale, Florida, 3–6 June 2001 reproduced in *J. Telemed. and e-Health*, Vol. 8, no. 2, 2002, p. 137.

3

Information Technology Empowering Differently Abled Persons

Ms. Chhavi

Lecturer Department of Engineering & Technology, MOU, Rohtak

Abstract

Information and Communication Technology (ICT) opens up great opportunities to improve the quality of life of differently abled persons. However, without substantial social effort, there is a risk that these technical developments will only give us products and services which increase the information gap. If determined efforts are made, ICT can indeed become an effective tool allowing a greater number of people to participate in society.

Disability is not a tragedy but an inconvenience. About 600 million persons or one-tenth of the world population is estimated have disability in some form or the other–whether visual, auditory, physical, speech-related, cognitive or neurological. I CT fundamentally aims to meet our demand for information and knowledge and our demand to communicate the same. Technology that facilitates meeting this basic human need is therefore very useful. In introducing an ICT-based methodology, there is the need to consider how to help differently abled persons gather and disseminate knowledge and information. In this regard, certain questions arise: what information to convey? Which sources to access? How ICT? Which ICT? This paper focuses on these questions, particularly how ICT can be used for the collection of information, its documentation and dissemination to society for the benefit of differently abled persons. Finally, it looks at an old yet reinvented strategy for including the differently abled in society. This strategy is community-based rehabilitation.

Introduction

Information and Communication Technologies (ICT) offer remarkable opportunities for the cost-effective production and dissemination of information products tailored to the specific needs of the differently abled. So far, many sections of society have been oriented to ICT. However, the major section that is left out includes differently abled people. This inequality needs attention.

It has been recognized that globalization and new information technologies are transforming the way the world is organized and information globally shared. These changes could accelerate progress in enabling the differently abled, but unless policymakers, practitioners and communities give attention to this marginalized group when considering the opportunities and risks inherent in ICT, and unless people themselves have a voice in how these new technologies are developed and deployed, these new technologies could very well exacerbate existing inequalities. Thus there is an urgent need to take advantage of the ICT revolution and ensure that it benefits all sections of society.

Accessibility is Essential

People with disabilities want to be independent; they want to do things for themselves by themselves. This is a fundamental issue of human dignity, which is enshrined and enacted in good corporate responsibility and legislation and also enabled as good business practice. Good 'accessible' ICT systems can open up new possibilities and opportunities for people with disabilities, because they have built in facilities that enable such people to use systems independently. ICT systems which do not build-in these 'accessibility' factors will cause enormous frustration because they cannot be simply used by differently-abled persons on their own, as independent human beings. Differently abled persons will often need assistive technology, such as screen readers or modified mouse. These add-ons help but are rarely a complete solution. Systems work best when they are specifically designed for able and differently abled persons, using and positively supporting accessible technologies. Most ICT systems and websites are not fully accessible. This is not because the designers have willfully discriminated against people with disabilities, but because accessibility does not happen automatically; it needs the active support of all levels of management and ICT.

What are the Challenges

ICTs have created a border-free and barrier-less space for all and these technologies can be very well extended to differently-abled persons. Though different agencies are developing compatible ICTs for the differently abled, a large proportion of differently-abled persons are not able to access these technologies due to socioeconomic and cultural factors specific to the region. These factors can include

- Limited access to and availability of relevant information (content) for members' development needs
- Limited awareness about the potential of ICTs as a tool for information exchange and dissemination
- Limited skills that call for a lot of training and hands-on demos on the usage and application of ICTs
- Lack of access to ICT equipment and services. Products are yet to be designed to be functional and affordable for end-users, and to correspond to a need identified by differently-abled people themselves
- Products still not actually available to the end-user
- Products not high-quality, safe and reliable
- Lack of ICT skills. Most information is now available through new ICTs like the Internet as opposed to traditional ICTs, such as radio. However, this tends to exclude those that lack the skills to use computers.
- Technophobia, especially in handling new technology

How can we Overcome Challenges

For differently-abled persons, IT is in fact far more important than for others. The issue here is not of doing the same thing more quickly or in a simpler way with the aid of IT, but of being able to perform tasks independently, which would be impossible without IT. The following are some of the measures which need to be implemented:

— Enhance information and knowledge sharing for the differently abled by proposing mechanisms and tools through a participatory approach to generate, process and disseminate relevant local content with appropriate and acceptable ICT formats
— Train differently abled people in the new information technologies in order to foster and support their access to the labor market
— Inform and counsel differently abled people on how to use ICTs to establish new relations with their environment
— Create structures of telematic information and communication
— Detect market 'niches' offering possibilities for differently abled workers
— Help participants in the project get a job
— Produce training and dissemination materials
— Provide non-discriminatory legislation, social security systems and access to personal assistance and develop accessible infrastructure
— Conduct information-needs assessment
— Use existing information sources and mechanisms for linking creators/holders of knowledge with users
— Use Community based rehabilitation

Community Based Rehabilitation (CBR)

Community based rehabilitation or CBR may be defined, according to three United Nation Agencies, ILO, UNESCO, and the WHO, as 'a strategy within community development for the rehabilitation, equalization of opportunities, and social integration of all people with disabilities. CBR is implemented through the combined efforts of differently abled people themselves, their families and communities, and the appropriate health, education, vocational and social services' (WHO, 1994). It differs from independent living (IL) in that, according to Lysack (1994), the entire community is the target of CBR programs; the CBR model is one of community development or partnership while IL ideology places control squarely with differently abled consumers.

'CBR is a strategy that calls for the full and co-coordinated involvement of all levels of society: community, intermediate and national. CBR as concept seems to be very naïve and realistic, but its implementation involves the co-operation of various members of society be it at community level or at an individual level. The defining quality of CBR is the involvement of communities'. Together with this CBR also requires the full fledged support of national policy-makers, a management structure, and the support of different government ministries, NGOs and other stakeholders (multi-sectoral collaboration). The CBR an approach goes beyond narrow concept of 'rehabilitation'. It does not require that differently abled people have to travel to a remote centre or a remote centre or institution to meet their needs. It requires support of multiple sectors to make a holistic impact on society for inclusion of differently abled persons. Being a long-term strategy CBR involves capacity building of differently abled people and their families, in the context of their community and culture. Its essential feature is its focus on partnership and community participation. This can help integrate the person into the community, a community which values the unique contribution which the person is able to make. CBR is not an approach which only focuses on the physical or medical needs of a person but a service that treats differently abled people as active recipients who can participate fully in the activities of the community.

Despite the fact that CBR has been in place for over two decades, some of its aspects need to be re-addressed. As the world is changing at a fast pace, new economic, social and medical models of CBR need to be developed so as to achieve sustainability. Also in line is the expansion and up-scaling of CBR programmes that will further include people from all age groups and tackle the issue of gender inequality. The implementation of CBR requires formal training to ensure effective management of programmes, meaningful participation of community people and satisfactory delivery of services from CBR workers and professionals who provide referral or support services. Nevertheless it has always been perceived that CBR as a strategy still has a long way to go for the inclusion of differently abled people in society.

Conclusion

The phenomenal growth of ICT-development has created a digital divide not only between the developed and developing countries, between rural and urban societies, and within the developing countries, but also among different segment of larger social groups, and the serious casualties here are the differently abled. ICT is a revolution and therefore it is inevitable as well as necessary to find possible options of ICT-use to socially include people with disabilities and to investigate uses of ICT and their devices to effect social inclusion.

Therefore, there is a need for research that seeks to identify differently abled groups, which are often targeted in an ICT-based social inclusion policy. Also we should study how these differently abled people can use ICT as a tool for their education/learning, business/work in everyday activities.

Consultations with several organisations has led to a consensus that a network should be formed and upon formation, should establish a companion website on which to profile the work of organizations working with the differently abled as well host an electronic mailing list to facilitate information-sharing and dissemination. This will fulfill the need for information-sharing and dissemination by capitalizing on the opportunities available through the Internet. Such websites and mailing lists are key sources of information about and for organizations dealing with the differently abled, and are key resources for members and interested partners.

References

[1] Bhalla, A. and F. Lapeyre, 'Social exclusion: Towards an Analytical and Operational Framework', *Development and Change* 28(3), 1997, pp. 413–433.
[2] Bijker, W.E. and J. Law, eds. *Shaping Technology/Building Society. Studies in Sociotechnical Change*, Cambridge: MIT Press, 1992.
[3] Burchardt, T., 'The dynamics of Being Disabled', *Journal of Social Policy* 29(4): 2000, pp. 645–668.
[4] Fisher, K.T. and P.B. Urich, 'Information dissemination and communication in stakeholder participation: the Bohol-Cebu water supply project', *Asia-Pacific Viewpoint*, Vol. 4, no. 3, 1999, pp. 251–269.
[5] 'The Use of Information & Communication Technology (ICT) to Support Independent Living for Older and Disabled People', Report by UK Dept. of Health, October 2002.
[6] 'The components of community-based rehabilitation programmes' accesed at http://www.unescap.org/esid/psis/disability/decade/publications/cbr.asp#sector

4

Barrier-free Higher Education for the Visually Challenged

D. Ramkumar

India is a country with more than 400 universities and around 18,000 colleges, including deemed-to-be universities and autonomous colleges aspiring to be centres of excellence. Ever since the Eighth Five Year Plan, the Ministry of Human Resource Development (MHRD) in consultation with the University Grants Commission (UGC) has taken several initiatives to improve and standardize the quality of higher education in the country. At present we are in the process of implementing the Eleventh Five Year Plan and it is the need of the hour to assess whether the initiatives taken for higher education reach people from all walks of life. Though the government and the UGC take initiatives to strengthen higher education in India, the responsibility of implementation is vested upon all central and state universities along with their various affiliated colleges. Any institution can attain standards of excellence only when it is capable of giving qualitative higher education to the less-privileged and the disadvantaged aspirants of our country in compliance with the guidelines issued by the UGC from time to time.

Here among the less-privileged and the disadvantaged, I would like to focus attention on the initiatives to be taken by the educational institutions to accommodate the special educational needs of the visually challenged aspirants.

Disability Rehabilitation Management through ICT, 27–30.
© 2012 *River Publishers. All rights reserved.*

Compared to any other disability, visual disability stands as a thread for a person's quest for higher education. There are special institutions of higher education for people with hearing impairment in our country through which they get qualitative education. Physically challenged aspirants also do not need any special educational arrangements except easily accessible classrooms, ramps on the steps, and so on. However, there is no higher educational institution exclusively for the visually challenged anywhere in India. Hence the only possibility is to accommodate them in mainstream institutions. This is in a way good since they will not be isolated from the mainstream society. In order to ensure the admission of visually challenged into higher educational institutions, the Government of India has introduced 1 per cent reservation for them within the total 3 per cent reservation for Persons with Disabilities (PWD). In spite of this, several higher educational institutions are hesitant to admit those people citing various reasons such as lack of specially skilled teachers and infrastructure facilities. Though some government institutions do admit them for the sake of implementing the quota for the disabled, lack of infrastructure and attention on the part of teachers forces many of the visually challenged students to discontinue their education.

Today there are only a handful of educational institutions that have the required facilities and that are catering to the special educational needs of the visually challenged. If other higher educational institutions too come forward with a positive approach towards providing higher education for the visually challenged, nothing will be impossible. As technological aid for the visually challenged are advanced in the international level, giving qualitative higher education on par with institutions in the West is not an impossible task.

Now there are specially designed software that convert printed textbooks into voice, that enable the visually challenged to type their documents and even browse the Internet independently. The only problem is that these software are expensive. But the visually challenged cannot be denied their right to higher education merely because of the cost factor. Hence higher educational institutions should take maximum care to put infrastructure facilities in place by availing the special fund provision available with the UGC for the purpose.

The UGC from time to time introduces special provisions and guidelines to make campuses disabled friendly. On 19 January 2009, it released a consolidated report on various provisions for the disabled to be strictly implemented in all higher educational institutions. The report says that each institute should immediately establish a disability council to coordinate the disability-related issues without any delay under the supervision of a faculty member and two assistants who will be paid by the UGC. The council would take care of infrastructure facilities, placement opportunities and arrangement of scribes for the visually challenged during examination. In this way, half the burden is met by the UGC. At present, this council is available only in a few Central Universities like the Jawaharlal Nehru University, New Delhi, Delhi University, Hyderabad University, the English and Foreign Languages University, Hyderabad, and autonomous institutions like the Tata Institute of Social Sciences, Mumbai and Loyola College, Chennai. The Pondicherry University is likely to have a special reading library for the visually challenged soon with advanced assistive technology.

When we discuss the process of teaching, teachers having visually challenged students in their class should pay extra attention to ensure that everything they teach reaches the visually challenged. For instance, when they use the blackboard, they should also speak out what they are writing. If necessary, they should be ready to give some extra time to the visually challenged outside the classroom.

Most teachers in higher educational institutions are unable to cater the needs of the visually challenged students as they are unaware of the methods to be followed and they very rarely come across such special students. Hence, teachers are not entirely to be blamed. One possibility is that the UGC arrange for a special orientation for all university and college teachers through the Academic Staff Colleges functioning in various universities.

Another major problem faced by the visually challenged students pursuing higher education is during their examination. No matter how good a student is in academics, if he does not get a good scribe, all his efforts will be in vain. At the same hand there is every possibility that academically dull students gain an upper hand thanks to a highly competent scribe. Hence, the real capability of a visually challenged student cannot be assessed through written exams including entrance exams for admission into various courses.

There are two ways in which this problem can be rectified. The first method is to orally record the student's answers on a tape or alternatively make the students type the answers on the computer with specially designed talking software. This will also save the educational institutions the burden of hunting for scribes.

It is not enough for the visually challenged students are educated on par with others. They should be given equal opportunity for employment in institutions of higher education as per the rule in force for persons with disabilities. Higher educational institutions should pay attention to ensure the visually challenged mingle with mainstream educated.

The use of terms like 'differently abled' instead of disabled and 'visually challenged' instead of blind alone will not give social security to. Real social security will come from rehabilitation effected through qualitative higher education and suitable employment.

India has 25 per cent of the world's visually challenged people. The literacy rate among 45 million visually challenged people is a mere 3 per cent. In spite of the desire for higher education, even the pursuance of graduation is beyond the reach of most visually challenged aspirants. Very rarely do the visually challenged enrol for research degrees and even those who do are often unable to complete it successfully owing to various factors. Less than .05 per cent of the visually challenged people are able to get suitable employment opportunities. The rest suffer in the doom of darkness.

This adverse situation can only be changed with the support of higher educational institutions and universities. To fulfill this task, the involvement of the corporate sector, IT industry and premier technological institutions like the Indian Institutes of Technology and Indian Institutes of Information Technology is very essential. The IT industry and various IITs and IIITs have made commendable and enduring efforts for the empowerment of the visually challenged through various innovative projects, special employment training and placement, awareness programmes and various other beneficial activities.

Part II

Infrastructure and potentials for use of ICT in Disability Rehabilitation

5

Role of ICT in Disability Rehabilitation in the Rural Environment

Dr. Ganesh Arun Joshi

Asstt. Prof. in PMR/Medical Edu., Composite Regional Centre for Persons with Disabilities, Punarwas Bhawan, Khajuri Kalan, Piplani, Bhopal-462021

The Problem

At present the services available in the developed countries are due to development in all spheres of the society, viz., medical sciences, engineering, information technology, literacy, awareness and paying capacity with due importance to research and development. The disabled people are achieving independent living due to social awareness and minimization of physical barriers.

The basic preventive and therapeutic services are not well developed in India till date. The scenario of rehabilitation is still worse. Although India has good rehabilitation expertise, it is restricted to a few major urban areas. Most of the rehabilitation prescriptions remain on paper due to problems in feasibility and availability of manpower and devices. The real-life situation for disabled persons in rural India is miserable.

Starting from spadework of awareness to the techno-savvy human-machine interface of latest assistive technology, there is a big gap in provision of rehabilitation services in rural India. The relative deficiency of manpower, uneven spread of services (urban concentration) and poverty-illiteracy combine slow the pace of rehabilitation services. The age-old welfare measures spring from the widespread belief of charity and infuse passivity in the disabled persons.

Disability Rehabilitation Management through ICT, 33–46.

Infant mortality is reducing at the cost of increasing birth-related impairments and disabilities like cerebral palsy and mental retardation. High consanguinity, environmental hazards, inadequate maternal care and nutrition have their own teratogenic effects. However, when fast developmental growth is achieved in the country, it falls short of reaching the needy due to poverty and population explosion. A number of spinal injury patients don't get rehabilitation facilities in our country and have to succumb to its complications. There are amputees who live their life with the help of crutches or manual support of caregivers in this country known for the world famous 'Jaipur foot'. There are only a few groups working on advocacy of the rights of the disabled persons. There is scarcity of parent organizations and self-help groups to support the rights of the disabled persons.

Presently orthopaedic, paediatric, and neurology specialists are forced to look into the rehabilitation aspects of various disabling conditions — a responsibility in which they are not trained. The scarcity of specialists in the field of rehabilitation adds to the problem. Even in hospitals having Physical Medicine and Rehabilitation (PMR) specialists, the referrals to the specialists are very few because fellow physicians in other fields don't know the utility and wide applicability of the PMR specialists. The present network of rehabilitation services in India is not well coordinated and is devoid of supervision by the PMR specialists. Hence the disabled persons become vulnerable to inappropriate rehabilitation management. Every year the district authorities try to provide disability certification to 100 per cent of the disabled population, but fail to do so in the absence of adequately trained staff, awareness, and other resources.

To give an example, a case study of spina bifida in north Chhattisgarh villages (Figure 5.1) shows grossly substandard services available. This child cannot stand or walk and hence her mobility is restricted. She needs special care for bedsores and has poor control on urine and stools. The ultimate kidney failure becomes a frequent cause of premature death in such cases. Who will take her to play, to school and what about her future career? Apart from being preventable to a large extent by care of maternal nutrition (folic acid supplementation), the condition calls for lifelong rehabilitation services.

Rural Environment

With nearly three-fourths of our country living in villages, it is the prime need that rehabilitation services are provided in rural areas. Rural society is a

Fig. 5.1 1 — A child with spina bifida having bedsore at the knee and unable to stand or walk. She has poor control of urine and stool (not obvious in the photo).

close-knit society with common culture and interpersonal communication. It has a lot of manpower with sufficient caregivers who are willing to take care of the basic needs of a disabled member of their society. The rural areas are full of resources (farming land, cottage industries) to facilitate independent living as well as vocational placement for the disabled. Poverty, illiteracy and infrastructure deficits are obvious drawbacks that stall the spread of rehabilitation services. In addition, there are superstitions and quacks that drain off the finances, time and psychological energy of the disabled persons and their care givers. The rural people do a lot of doctor shopping with non-realistic expectations and goals set by the local healers.

However, in the recent years the rural scenario is changing due to better infrastructure development, reach of medical services, penetrating telecom and satellite communication, improved literacy, and awareness generation. The rural-urban migration has also reduced over the last decade. Low expectations from the disabled in the rural society give them lifetime support them but encourage passivity in them. This blocks the principle of making the disabled person independent.

Disability

Disability is a link in the chain which starts from a disease process. Some diseases like common cold get cured without any lifelong marks but there are many diseases that cause permanent impairment. The pertinent definitions are as follows:

- *Impairment* is any loss or abnormality of psychological, physiological or anatomical structure or function.
- *Disability* is any restriction or lack (resulting from an impairment) of ability to perform an activity in the manner or within the range considered normal for a human being.
- *Handicap* is the disadvantage for a given individual resulting from an impairment or disability that limits or prevents their fulfillment of a role that is normal depending on age, gender and social and cultural factors for that individual.

Census 2001 data shows an overall prevalence of 2.33 per cent of disability in India with 16.4 million disabled persons living in the rural areas (Table 5.1). The male to female ratio is 4:3.

The disabled persons are categorized into two main categories, viz., physical and mental (Figure 5.2).

Table 5.1. Number of disabled persons in India (in lakhs)

	Rural	Urban	Total
Males	94.10	31.95	126.05
Females	69.78	23.23	93.01
Total	163.88	55.18	219.06

Source: Census 2001.

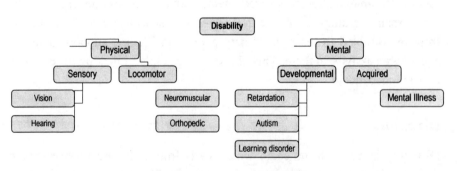

Fig. 5.2 Major types of disability.

The locomotor disabilities result from impairments in the limbs, for example, paralysis, amputation, deformities, and so on. The visual disabilities may be either blindness with total absence of sight or low vision with gross diminution of vision that cannot be corrected by medicine, surgery or use of spectacles. The hearing disabilities result from varying degrees of hearing deficit. The hearing deficit present right from birth implicates impaired speech development culminating in deaf-mutism. Mental retardation is a condition characterized by low intelligence and adaptive behaviour deficits right from birth. Autism is another condition affecting child development where the communication and socialization of the child are grossly affected. Learning disorders are so frequent in classroom students that it is not considered as a disability by many but is typified by difficulty in acquiring information/studies due to disordered processing of sensory signals (visual, hearing, and so on) in the brain. Mental illness is an acquired condition as a result of disease of the thought process with intelligence remaining unaffected.

Persons with Disabilities Act, 1995

The Government of India implemented this act in February 2006. The act defines seven categories (Table 5.2) of disabilities to be served and is further in process of amendment for addition of more disabilities. According to the act, a person suffering from not less than 4 per cent disability as certified by a medical authority is eligible for certain benefits to facilitate education, vocation and independent living.

Rehabilitation

As for any human being, the disabled persons have their strengths and weaknesses and suffer similar health problems. The disabling conditions are not diseases that can be cured but rather need rehabilitation services lifelong.

Rehabilitation is defined as the development or restoration of a person to the fullest physical, psychological, social, educational, vocational and avocational potential consistent with his/her physiological or anatomic impairments and environmental limitations. Rehabilitation is a problem-based approach with a goal to attain independence in activity and improve quality of life.

Table 5.2. Categories of disability as defined by PDA, 1995.

S.N.	Category	Definition
1	Blindness	A condition where a person suffers from total absence of sight or visual acuity not exceeding 6/60 in the better eye with correcting lenses or limitation in the field of vision subtending an angle of 20° or worse
2	Low vision	A person with impairment of visual functioning even after treatment or standard refractive correction but who uses or is potentially capable of using vision for planning or execution of a task with appropriate assistive devices
3	Leprosy cured	Any person who has been cured of leprosy but is suffering from loss of sensation, paresis or deformity due to leprosy
4	Hearing impairment	Loss of 60 dB in better ear in the conversational range of frequencies
5	Locomotor disability	Disability of bones, joints or muscles leading to substantial restriction of movement of the limbs or any form of cerebral palsy
6	Mental retardation	A condition of arrested or incomplete development of mind of a person which is specifically characterized by subnormality of intelligence
7	Mental illness	Any mental disorder other than mental retardation

Rehabilitation involves a twopronged attack on the disabled person as well as the environment.

Rehabilitation strategies for the disabled person include reversal or reduction of impairment; substitution of lost function by using the person's residual abilities and/or use of devices to assist the person to overcome the disabilities. Environmental actions include awareness creation, identifying resources in the local environment to facilitate independence and productivity, social reforms and modification of physical barriers for persons with disabilities.

The key to success of rehabilitation is good teamwork in accordance with the realistic goals worked out with active participation of the disabled person. A team (Figure 5.3) of rehabilitation specialists is led by the PMR specialist. The physiotherapist, occupational therapist, prosthetic and orthotic engineer, speech pathologist, clinical psychologist, rehabilitation nurse, medico-social worker, and other necessary members complete the team. The diagnosis and rehabilitation planning is done by the PMR specialist in a multidisciplinary team. The necessary therapy is provided by relevant rehabilitation professionals. Assistive devices are provided by the P&O engineer and assistive technology provider. The special educator helps to sort out the educational problems. The vocational counselor with the trainer, occupational therapist and medico-social worker help the disabled person to return to a gainful vocation.

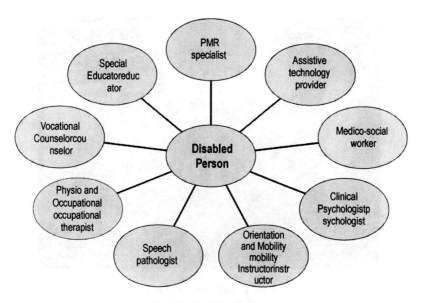

Fig. 5.3 Rehabilitation team.

The clinical psychologist helps the disabled person to come to terms with the coping process of disability and provides motivation to participate in the rehabilitation process for the best possible outcome. Sports and recreation by the recreation therapist provides the means for existence, supplements therapeutic programme, builds up endurance and lowers incidence of medical complications. In some institutes, the rehabilitation process is characterized by close communication between the members of the team; such teams are called interdisciplinary teams. Sometimes the therapeutic work on the patient is anchored by a single professional; the team in such a scenario is called a trans-disciplinary team.

Institution- Vs. Community-Based Rehabilitation

Well-established rehabilitation centres are available in private, public and government sectors in select major cities of India. These centres provide services in major institutional set-ups where the disabled persons have to commute from their home. Some of them provide outreach services in remote rural areas through camps (Figure 5.4). Thus, rehabilitation services are available in villages only once in a blue moon. Disability identification, certification,

Fig. 5.4 Audiometry in a Camp.

prevention, intervention, therapy, assistive devices, accessibility, education, vocation and recreation all need a modified approach in rural areas. The popular tricycle also fails in hilly/muddy areas of the rural environment. Thus, wheelchairs, artificial limbs, hearing aids, and remote-operated environment control devices all are destined to fail unless the rural environment and its infrastructure are studied and modifications in the approach are done.

The other mode of providing rehabilitation facilities is through community-based rehabilitation (CBR) approach. This approach utilizes local resources (manpower and raw material) to rehabilitate disabled persons of the locality (Figure 5.5). CBR provides a democratic approach to rehabilitation by training the local expertise to use local resources for therapy, assistive devices, education and vocational avenues. CBR suites the community in the best way and has active involvement of the community to run a self-maintained rehabilitation process.

Assistive Technology

Activities of Daily Living (ADL) are the basic activities of day-to-day living which include mobility, self-care, communication, management of

Fig. 5.5 Mobility training to a blind lady in a CBR setting.

environmental hardware and sexual expression. They form the basis on which the assistive devices are designed. Animate assistance is provided by caregivers, pets (e.g. guide-dog), inanimate assistance can be provided by mechanical or electronic devices. With the advent of varied types of HMI (human-machine interfaces) and progress in ICT, assistive devices have found new, expanding horizons. The devices are working like magic wand and facilitating communication and activity.

In the era of automobiles, domestic animals for transport are being neglected due to slow speed, but they seem to hold promise for the transport of disabled persons in the present energy crisis. These trained animals can be the best friends for disabled persons, e.g. guide dogs for visually impaired persons. The road network is improving and seems to hold promise for supporting motor transport and use of wheel-based mobility.

The discipline of using devices to improve function is known as assistive technology. An assistive device may be any item, piece of equipment or product system, whether acquired commercially off the shelf, modified or customized, that is used to increase, maintain or improve the functional capabilities of disabled persons. The burden of the caregiver is reduced by use of assistive devices. The conventional devices include artificial limbs (prostheses; Figure 5.6), calipers and splints (orthoses), mobility aids (canes, crutches, walkers and wheelchairs; Figure 5.7) and modified vehicles (tricycles, scooters, cars, and so on, with necessary modifications). With advanced technology, mechanical hoists, stair-lifts (Figure 5.8), elevators and a lot of customized modifications are available for the most severely disabled persons. Modified sports equipment for the disabled is adding to the recreational facet of rehabilitation.

Persons with visual impairment make use of Braille–a script with letters written in combination of six dots embossed on thick sheets that can be read by touch. The sense of touch, smell and hearing is maximally exploited to replace mobility in them. Thus, typical bad smell indicates where the toilet is, recognizing voice of a colleague in a office suggests his chamber and an embossed pictograph with landmarks along the floor and walls guides the person to a particular place. For persons with low vision, improved utilization of vision can be effected with use of various optical/non-optical devices. There are various hardware and software that can be used for the disabled. The hearing impaired

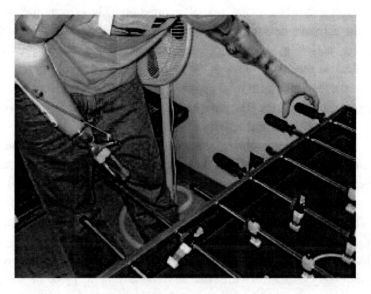

Fig. 5.6 Upper limb prostheses under use for recreation.

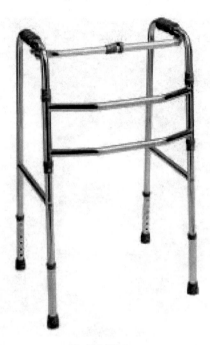

Fig. 5.7 Walker.

persons are able to hear with hearing aids. The cochlear implant (artificial ear) is an array of hardware that is surgically implanted in contact with the auditory nerve. An external receiver directly sends sound signals electromagnetically to this implant bypassing the ear. Electromagnetic loop induction system in training halls helps the hearing disabled persons to catch clear sound with the help of their hearing aids. The locomotor impaired persons need devices for mobility, feeding, toileting, dressing, writing, and so on.

A wide spectrum of assistive devices powered by computer-based adaptations (Figure 5.9) include environmental control units (alarms, remote switches), communication devices (internet, augmentative and alternate communication), educational devices (Braille, hearing aids, multimedia), therapeutic tools (biofeedback, occupational training), job facilitation (desktop office, graphics, CAD/CAM), recreation facilities (modified sports equipment), artificial limb controls (myoelectric hand, intelligent prosthesis) and ever expanding horizons with neuroprostheses (artificial ear, functional neuromuscular stimulation and research with artificial eye). Robotics and ICT with a wide spectrum of HMI and switch controls like those operated by finger,

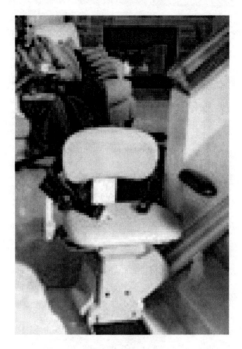

Fig. 5.8 Stair-lift.

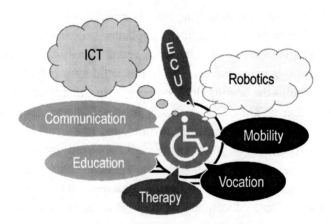

Fig. 5.9 Computer-based assistive technology.

body, tongue, touch, bioelectricity, sip-and-puff, blink, eye gaze, sound/voice, and so on provide opportunity for disabled persons in virtually all categories and grades of disability to achieve a better independence level.

**Role of Information Communication Technology (ICT)
in Rehabilitation**

The Ministry of Social Justice and Empowerment, Government of India, has a network of national institutes, regional centres and district centres (DDRCs (District Disabilities Rehabilitatuion Centers)) for the purpose of rehabilitation services. They work with help of NGOs and private agencies in the far fetched areas of the country. The Rehabilitation Council of India (RCI) is a statutory body for standardization of human resource development in rehabilitation professions. The government runs various schemes to provide assistive devices and rehabilitation services to the disabled persons.

ICT has a role in interventions on both the fronts of rehabilitation, viz., the disabled person and his environment. The primary health care and rehabilitation follow-up for the disabled person can be made available through telemedicine. The assessment and marketing of assistive devices can be facilitated by ICT. Human resource development for care of the disabled persons is facilitated by satellite downlink programme initiated by the RCI in association with the Indira Gandhi National Open University (IGNOU). Survey and medical record management can be done and shared with the help of ICT.

ICT has potential for awareness creation to improve knowledge, attitude and practice, thus helping remove social and physical barriers in the rural environment. The rehabilitation institutions in India have their own way of providing services with poor communication amongst them and hence lack facilities and skills that can be called wholesome. ICT can help exchange of ideas amongst various institutes and CBR providers to achieve optimal outcome of rehabilitation. Disability needs survey, certification, therapy, telemedicine-based treatment procedures have a potential to supplement CBR.

With the target of Community Service Centers in one lakh villages in India by the end of December 2008, the ICT infrastructure will be in place in India. Feasibility issues in rural environment are important considering the infrastructure problems of road, electricity, water, and telecommunications. Alternative sources of energy like biogas, solar energy, hydropower and wind power making the villages self-sufficient is essential to speed up the progress.

Rural India has a primary health care service that needs to be sensitized to the issues of disabled persons. Surgical facilities in far away districts have been successfully provided by the LifeLine Express. Why not provide one Lifeline Express for each state to handle the huge Indian population?

Awareness and training can be accomplished with the help of ICT. Assistive devices, fitment maintenance and rehabilitation facilities can be provided through the existing Primary Healthcare Centres (PHCs) with guidance from DDRCs. Education to the disabled persons can be provided through general schools by inclusive education or by special schools. For the severely disabled, home-based education shall be provided by distance mode or home visits as being done by the Rajya Shiksha Kendra, Madhya Pradesh. The issues on vocation, recreation and access of the disabled persons shall be discussed in the panchayats and at the *chaupal* and will get strength from e-governance.

Awareness and advocacy help in improving the social security of the disabled persons. The standard approach through agencies like PHCs, *anganwadis*, panchayats, police, schools, post office, railways, and so on will be required. The widespread infrastructure of the post offices can be well exploited to support ICT facilities throughout rural India. Incentives and awards for novel rehabilitation practices and devices can spark off the sleeping R&D sector in disability rehabilitation. Combined projects of CBR with the help of ICT are the need of the hour to facilitate the reach of rehabilitation facilities to the majority of Indians. Public-private partnership and third party payers are in place to provide necessary funding.

Summary

Summarizing, it is stated that the strengths of a disabled persons need to be built upon with the help of ICT. Rather than the healthy population who are facing the wrath of sedentary life due to ICT, the new developments shall target the disabled persons for applications of different types of software and hardware to provide appropriate assistive devices. The underlying effort throughout should be to break the vicious cycle of poverty, disability, segregation, powerlessness and charity that leads to the denial of participation, respect and opportunities for persons with disabilities in the society. The rehabilitation team approach, assistive devices, role of information technology and agents providing the rehabilitation services will be presented in this article giving a statement of needs and management strategies for the rural disabled population.

6

National Programme for Control of Blindness in India

A Rathore

Demographic Profile

The Registrar General of India has estimated that the population of India is 1,095.70 million [2005] and projected to increase to 1,254 million by 2015 with the proportion of children in the age group of up to 15 years and elderly (60 years and above) constituting 33.50 per cent and 7.20 per cent of the total population respectively [1]. Similarly, infant mortality rate per thousand live births has declined from 74 in 1995 to 57 in 2007 [2], which has resulted in saving larger proportion of low-birth weight and premature babies largely due to simultaneous development of neonatal and child care services in the country. However, these children are at a greater risk of developing refractive errors such as high myopia, myopic astigmatism, anisometropic amblyopia and strabismus, and retinopathy of prematurity [ROP] [3].

At the other end of the age spectrum, with increase in life expectancy, there has been a sharp increase in the number of elderly persons between the years 1991 and 2001 and this figure is projected to rise to about 324 million by 2050 [4]. The elderly are prone to cataract and other age-related ocular morbidity. Emerging eye issues like diabetic retinopathy, glaucoma, low vision and childhood blindness need attention and multipronged solutions. In addition, all age groups and specific category of workforce are vulnerable

Disability Rehabilitation Management through ICT, 47–56.
© 2012 *River Publishers. All rights reserved.*

to ocular trauma. Ocular morbidity hampers performance at school, reduces employability and productivity and generally impairs quality of life, which has a direct bearing on the economic health of the nation in terms of GDP.

There is a huge unmet need for eye care throughout the world and this is increasingly being recognized as a vital component of the total health care delivery system.

Global Perspective on Blindness

In 1988, the number of people who were blind [visual acuity (VA) <3/60 in the better eye] was estimated to be 37 million worldwide. By 2002–04, the latest period for which we have data, it was estimated to be 45 million: 8 million due to uncorrected refractive errors and 37 million blind due to other causes [5–7]. Over the last 20 years, the causes of blindness have changed in proportion and actual number. Cataract has remained the major cause of blindness globally, more so in Asia. The proportion of cases of blindness due to some of the major causes are given in Table 6.1.

The number of people blinded by trachoma, onchocerciasis and vitamin A deficiency have tended to decrease over the last 20 years [8].

National Programme for Control of Blindness

The first organized effort towards addressing blindness on a national scale was undertaken by the Government of India in 1963 by launching the National Programme for Trachoma Control. This programme was subsequently merged with the National Programme for Control of Visual Impairment and Prevention

Table 6.1.   Leading Causes of Blindness and Proportion of People Affected by Each Cause.

Cause	Number and proportion of population affected
Cataract	39% (17.60 million)
Refractive errors	18.80% (8 million)
Glaucoma	10% (4.50 million)
AMD	7% (3.20 million)
Corneal scar	4% (1.90 million)
Diabetic retinopathy	4% (1.80 million)
Childhood	3% (1.40 million)
Trachoma	3% (1.30 million)
Onchocerciasis	0.70% (0.30 million)
Other causes	11% (4.80 million)

Table 6.2. Global Estimate of the Number of People with Visual Impairment (blindness and low vision).

Definition	No. of people [million]
Blindness [eye disease] <3/60 to no light perception	37
Blindness [refractive error] <3/60 to light perception	8
Blindness [all causes]	45
Low vision [eye disease] <6/18 to 3/60	124
Low vision [refractive error] <6/18 to 3/60	145
Low vision [all causes]	269
Total visual impairment [all causes]	314

of Blindness which in 1976 was redesigned as the National Programme for Control of Blindness (NPCB). The programme was launched as a Central Government-sponsored scheme with the goal of reducing the prevalence of blindness from 1.40 per cent to 0.30 per cent by the year 2000. However, as the country was unable to achieve the stated target in the set time frame, it was revisited and a new time frame was established through various consultative processes and discussions. The revised dates were realigned for the year 2020 under the aegis of 'VISION 2020: The Right to Sight' initiative of the World Health Organization (WHO), the International Agency for Prevention of Blindness and various non-governmental organizations. See Table 6.2 for Global Estimate of the Number of People with Visual Impairment (blindness and low vision). As per a survey in 2001–02, the prevalence of blindness in India was estimated to be 1.10 per cent. The main causes of blindness in India are shown in Figure 6.1.

The objectives of the NPCB are to:

- Reduce the backlog of blindness through identification and treatment of blind, especially cataract.
- Develop eye care facilities in every district of the country.
- Develop human resources for providing eye care services.
- Improve quality of service delivery.
- Secure participation of voluntary organizations in eye care.
- Enhance community awareness on issues related to blindness.

The activities of the NPCB include:

1. Identification and treatment of cataract
2. Detection and correction of refractive errors including eye screening camps in schools

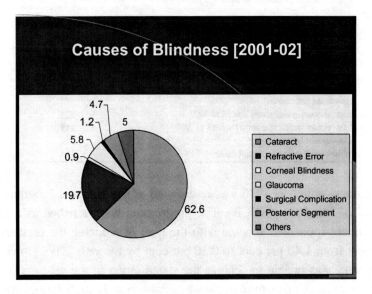

Fig. 6.1

3. Eye screening of persons in blind schools
4. Community Awareness (IEC)
5. Eye banking and cornea collection
6. Procurement of equipment and consumables
7. Infrastructure development and technological advancement
8. Training of health personnel
9. Tele-ophthalmology
10. Maintenance of equipment
11. Monitoring and evaluation
12. Mobilization of eye surgeons for cataract surgery drive, especially for the north-eastern states
13. Surveillance

Organization and Delivery of Eye Care Services in the Country

Organization

At the central level, the Secretary, Health and Family Welfare, Government of India, is the administrative head who is supported by officers of the ranks of Joint Secretary, Deputy Secretary and Under Secretary to the Government of India. The technical division is headed by the Director General (DG) of Health Services supported by the Additional DG, Assistant DG and Deputy Assistant DG. At the state level too the administrative and technical heads are structured on similar lines, however with different nomenclature, and assisted by various administrative staff. At the state level, the State Programme Officer [SPO (NPCB)] is the main focal person to coordinate activities in the state. At the district level, District Health Societies have been formed under the chairmanship of the District Collector for planning and implementing all blindness control activities. The Chief Medical Officer (CMO), the District Ophthalmic Surgeon along with the District Programme Manager coordinates all the activities.

Service Delivery

Basic eye care services are provided by medical officers and paramedical ophthalmic assistant at the level of primary/community health centre. Services of eye surgeons are available at district hospitals, medical colleges, and institutes of national importance in the government and non-governmental sectors. The process of procurement of equipment, instrument and infrastructure development has been decentralized at the level of states/districts so as to improve the efficiency in the system and remove delays.

Budget

In the Eleventh Plan (2007–12) the NPCB has been sanctioned a sum of Rs. 1,250 crores for the entire country. The earmarked budget for states/Union Territories are released based on the their annual plan of action, their capacity for financial utilization, need, and furnishing of previous year's utilization certificate (UC), statement of expenditure (SOE), and audit statement. The requisite amount is released to the State Health Society either through electronic transfer or bank draft, which in turn releases it to respective District Health Societies. Allocation and expenditure of the NPCB are shown in Table 6.3.

Table 6.3.   Budget allocation and expenditure under NPCB (2004–05 to 2010–11).

Year	Budget Allocated [crores]	Expenditure [crores]
2004–05	88.00	87.20
2005–06	93.32	92.84
2006–07	111.87	111.53
2007–08	165.20	164.95
2008–09	250.00	249.50
2009–10	250.00	252.89
*2010–11	260	126.73

*As on December 2010

Table 6.4.   Performance of Cataract Surgery [lakhs].

Year	Target	Achievement	% IOL
2004–05	42,00,000	45,13,667	88
2005–06	45,13,000	49,05,619	90
2006–07	45,00,000	50,40,089	93
2007–08	50,00,000	54,04,406	94
2008–09	60,00,000	58,16,183	94
2009–10	60,00,000	59,06,016	95
*2010–11	60,00,000	27,00,000	95

*As on December, 2010.

Achievements

(A) *Cataract surgery:* Performance of cataract surgery has been steadily increasing, as indicated in Table 6.4. Of all the cataract surgeries performed, almost 94 per cent have been with intra-ocular [IOL] lens implant.

(B) *School eye screening programme:* One of the essential strategies of the NPCB is to ensure eye screening amongst children studying in government schools throughout the country. See Table 6.5 for Performance under school eye screening component.

(C) *Collection of donated eyes:* Hospital cornea retrieval programme is the main strategy for collection of donated eyes, which envisages motivation of relatives of terminally ill patients, accident victims and others with grave diseases to donate the eyes of their love ones. Eye donation awareness fortnight is organized from 25 August to 8 September every year to promote eye donation/eye banking. Gujarat, Tamil Nadu, Maharashtra, Delhi, Chandigarh, Andhra Pradesh, Kerala and Karnataka are the leading states in eye collection/cornea retrieval activity.

Table 6.5. Performance under school eye screening component.

Year	Teachers Trained	School Children Screened	Children Detected with Refractive Errors (%age)	Poor Children Provided Free Glasses (%age)
2004–05	97,310	2,68,62,932	5,72,691 (2.13)	2,83,070 (49.4)
2005–06	1,26,163	2,97,37,168	7,71,901 (2.6)	3,85,403 (49.9)
2006–07	2,03,221	3,54,29,289	9,63,168 (2.71)	4,56,634 (47.4)
2007–08	1,93,629	2,76,76,430	11,26,985 (4.07)	5,12,020 (45.4)
2008–09	1362383	3,08,50,532	11,29,046 (3.65)	5,12,000 (45.3)
2009–10	1085883	3,64,278	10,04,785 (3.27)	5,05843 (50)
*2010–11	38877	1,05,97,373	3,97,096 (3.74)	1,14,882 (30)

* As on December, 2010

Table 6.6. Number of eye balls collected between 2004–05 to 2010–11.

Year	Eye balls collected
2004–05	23,553
2005–06	28,007
2006–07	3,007
2007–08	38,596
2008–09	41,746
2009–10	46,589
*2010–11	20,915

* As on December, 2010

(D) *Training of ophthalmic surgeons:* Eye surgeons working in the public sector in different states/UTs are given training in various sub-specialties of ophthalmology in 30 select institutes in the government and public institutions and the expenditure is borne directly by the Government of India. The areas of training include ECCE/IOL, SICS, phaco-emulsification, low vision services, glaucoma management, paediatric ophthalmology, indirect ophthalmology, medical retina and laser technique, vitreo retinal surgery, eye banking and corneal transplantation, oculoplasty, and strabismus diagnosis and management. The training of other staff is planned, organized and implemented by state/Union Territory governments with financial assistance from the Government of India.

(E) Information Education and Communication (IEC) *activities:* IEC activities are undertaken at the central, state and district levels. Special campaigns for mass awareness are undertaken during the Eye Donation

Table 6.7. Eye Surgeons Trained (2004–5 to 2010–11).

Year	Eye Surgeons Trained
2004–2005	350
2005–2006	250
2006–2007	250
2007–2008	300
2008–09	450
2009-10	350
*2010–11	300

Fortnight (25 August to 8 September) and World Sight Day (second Thursday of October). An innovative strategy for community awareness on eye donation has been initiated by involving not only eye banks but also blood banks with the popularization of the slogan *'Jeete Jeete Rakt Daan, Mrityu ke Uprant Netra Daan' (Donate blood when alive, after death offer your eyes).* At the Central level, prototype IEC material is produced and disseminated to the states. Guidelines and training manuals are also prepared centrally and disseminated. A quarterly newsletter has also been started since July 2002.

(F) *Support to voluntary organizations*: Voluntary organizations/NGOs play an important role in implementing various activities under the programme. Under the National Rural Health Mission [NRHM] scheme, Blindness Societies have been merged with Health Societies, both at the state and district level. Under the scheme a non-recurring grant of a maximum of Rs. 30 lakhs is granted for expansion/upgradation of eye care units in tribal and backward rural areas. Another Rs. 15 lakhs for upgradation of eye banks and Rs 1 Lakh for upgradation and setting up Eye Donation Centre is granted as non-recurring assistance and Rs. 1,500 and 1000 is provided per pair of eyes as recurring assistance for registered eye banks and Eye Donation centres respectively. In addition, a sum of Rs. 50,000 is provided for setting up of vision centres. Under the tele-ophthalmology project, a non-recurring financial assistance of upto Rs. 60 lakhs is awarded based on the merit of the proposal, the area to be served, institutional capacity and favourable recommendation from the respective state/UT government.

New Initiatives Implemented Under Eleventh Five Year [2007–12] Plan Period

A budget of Rs. 1,250 crores has been earmarked for NPCB in the Eleventh Five Year Plan period. This includes continuation of the Tenth Plan period activities and proposed new activities:

- Construction of dedicated eye wards and eye operation theatres in district and sub-district hospitals in the north-eastern states, Bihar, Jharkhand, Jammu and Kashmir, Himachal Pradesh, Uttaranchal and a few other states as per demand.
- Appointment of Ophthalmic Surgeons and Ophthalmic Assistants in new districts in district hospitals and sub-district hospitals.
- Appointment of Ophthalmic Assistants in PHCs/Vision Centers where there are none (at present ophthalmic assistants are available in block-level PHCs only).
- Appointment of eye donation counsellors on contract basis in Eye Banks under the government and NGOs.
- Grant-in-aid for NGOs for management of eye diseases like diabetic retinopathy, glaucoma management, laser techniques, corneal transplantation, vitreoretinal surgery and treatment of childhood blindness. Financial assistance is given at the rate of Rs. 750 for cataract surgery and Rs. 1,000 per case of other major eye diseases as described above.
- Special attention to clear cataract backlog and take care of other eye care ailments in the north-eastern states by organizing and mobilizing teams of ophthalmologists from Central Government- and NGO-run hospitals.
- Development of mobile ophthalmic units in hilly/under-served states and initiation of telemedicine in ophthalmology (Eye Care Management Information and Communication Network) on pilot basis.
- Provision of latest equipment to RIOs, medical colleges, district/sub-district hospitals, CHCs, Vision Centres.
- Strengthening of Low Vision Services by providing training for health personnel and low vision aid to Regional Institutes of Ophthalmology (RIO) and a few select medical colleges.

- Provision of exclusive budget for maintenance of ophthalmic equipment.
- Involvement of private practitioners at the sub-district level.

References

[1] *National Health Profile*, Central Bureau of Health Intelligence, Directorate General of Health Services, Ministry of Health and Family Welfare, Nirman Bhawan, New Delhi, 2007.

[2] *SRS Bulletin*, Registrar General, October, vol. 43, no. 1, 2008, p. 5.

[3] Jalali, S., R. Anand, H. Kumar, M.R. Dogra, R. Azad, and L. Gopal, 'Programme Planning and Screening Strategy in Retinopathy of Prematurity', *Indian Journal of Ophthalmology*, 51, 2003, pp. 89–97

[4] Ingle, G. K. and A. Nath, 'Geriatric Health in India: Concerns and Solutions', *Indian Journal of Community Medicine*, vol. 33, 2008, pp. 214–18.

[5] S. Resnikoff, et al., 'Global Magnitude on Visual Impairment in the Year', *Bulletin of World Health Organisation* no. 82, 2002, pp. 844–51.

[6] S. Resnikoff, et al., 'Global Magnitude of Visual Impairment Caused by Uncorrected Refractive Errors in 2004', *Bulletin of World Health Organisation*, no. 86, 2008, pp. 63–70.

[7] International Agency for the Prevention of Blindness, 'State of the Worlds' Sight Vision 2020: The Right to Sight 1999–2005', London: IAPB, 2005.

[8] Foster, Allen, Clare Gilbert, and Gordon Johnson, 'Changing Patterns in Global Blindness: 1988–2008', *Community Eye Health Journal*, September, Vol. 21, no. 67, 2008, pp. 37–9.

7

Psychological Rehabilitation on ICT and Families of the Differently Abled

Neeradha Chandramohan

Director, National Institute for Multiple Disabilities, Chennai-603112, Tamil Nadu, India; niepmd@gmail.com

The closing two decades of the twentieth century witnessed significant development in the field of disability rehabilitation in general, with considerable focus on Information Communication Technology (ICT). Recent developments in ICT have revolutionized the lifestyle of persons with disabilities (PWDs) to a great extent. In developed countries, ICT has added quality and luxury to life but for developing and under-developed countries it caters to the basic needs of PWDs.

Utilizing technological advancement for the benefit of persons with special needs involves two fundamental dimensions. The first entails making use of technology to help them overcome the handicaps and limitations caused by the disabilities. The other consists of the adaptive and assistive techniques necessary to be installed and used in order to eliminate obstacles posed by various disabilities, in rendering the technology itself accessible to PWDs and their families.

The role of rehabilitation psychologists in this rehabilitation process lies in synchronizing the technology with available services delivery models for PWDs and their families. This role includes both direct services, such as

Disability Rehabilitation Management through ICT, 57–59.

psychotherapeutic interventions, applied behavioural modifications and social skills training, and indirect services, such as staff training and development and research planning and implementation.

Families play a significant role in the rehabilitation process of PWDs. Disabilities pose challenges to families at every subsequent stage of the life-cycle of the differently-abled individual, right from detection, screening, diagnosis, to intervention and rehabilitation. Many families do not have access to appropriate services due to non-availability of services or lack of awareness of the availability of such services.

Families that have been closely associated with persons with disabilities often undergo stress/psychic crisis and follow various models of adjustment to disability, ranging from shock, to denial, anger, disbelief, etc., prior to the adaptation process. Parents/families should thus be counseled at each phase, given emotional support and educated and updated about newly available rehabilitation services.

Perspectives of disability can be understood in terms of current concepts and in the context of international commitments. WHOs current standard and the International Classification of Functioning Disability and Health (ICF) emphasize that 'disabled people's functioning in a specific domain is an interactive process between their activities, health, condition and contextual factors'. Within this framework, technology in general and ICT's in particular are key interfaces.

Developments in technology, especially information technology, are dynamic in nature. They have immense potential to improve and manage different aspects of interacting with persons with disabilities such as adapted teaching, sharing of resources, professional development of teachers, increasing accessibility, research and development, human resource development, distance education, total quality management and bridging the gap between students with and without disabilities.

The application of universal design principles in mainstreaming ICTs as well as in assistive devices has a major role in improving the accessibility of living environment for persons with disabilities. As per a NSSO 2002 report, the significant population (58%) amongst persons with disability (1.8 per cent) group falls under the locomotor category. Yet, the majority of such patients do not have access to rehabilitative services as they live in remote, rural locations.

Alternatively, rehabilitation services may be made accessible to them with the help of technology, ICT in particular, with horizontal expansions of services.

ICT facilitates accessibility in various ways. It plays a vital role for registry through disability helplines and helps in early detection and prevention. The other areas ICT works in are automation of mobility aids, CAD/CAM and embedded systems of locomotor aids, Gait and Posture Analysis, robotics in assistive devices, Tele-rehabilitation, computer-assisted restorative surgery, education through videoconferencing/telecom and tele-certification.

However, impressive as the impact of technological development may be, the essential truth is that mechanical, electronic and other techniques can only assist in the rehabilitation/learning process of persons with disabilities, but cannot replace the human touch. Special educational needs can only be met by meaningful conversion of 'abstract philosophies and technological principles' into effective practice (Freiberg, 2000). The advancement of ICT in the field of disability rehabilitation may not be possible with mechanical manipulations alone; there is need for human involvement and convictions as well.

References

[1] Freiberg, L.K., *Educating Exceptional Children*, Guilford: Dushkin Publishers, 2000.

[2] Sansanwal, N., 'Information Technology and Higher Education', *University News — Association of Indian Universities*, 38(42), 16 October 2000.

[3] Mahopatra, C.S. ed., *Disability Management in India*.

[4] NSSO Report 2002.

[5] UN forum to examine how information technology can assist persons with disabilities, UN, Press Release, Department of Public Information, News and Media Division, New York PI /1766 /SOC / 4731, 26 March.

8

Vocational Rehabilitation of Persons with Disabilities in India

R. Narasimham*

*No. 5 Gokulam Colony, Off Shriniovas Pillai Estate, West Mamdalam,
Chennai-600033; rnsimham@yahoo.com, rnsimham@gmail.com*

Vocational rehabilitation is the most tangible and perceptible of all processes of rehabilitation. It even enables the person to purchase other rehabilitation aids. It also happens to be the last process that the parent or the client him/herself looks for. Given that the medical and surgical interventions take the front row and the educational process takes the next even among the non-disabled, it is no surprise that the profession of vocational rehabilitation itself is young, dating back to the end of the Second World War (1939–45). That is the reason why most of the rehabilitation professionals dealing with other professions such as therapy and education dabble in the field of vocational rehabilitation also.

The number of professional vocational rehabilitation workers is limited in India.

The field entered as a systematic process only in 1968 with the start of two Vocational Rehabilitation Centres for the Handicapped (VRC) at Mumbai and Hyderabad, as a Research project assisted by the USA. While vocational rehabilitation was one of the components in other hospital-based projects

*Consultant (Vocational Rehabilitation), Chennai

Disability Rehabilitation Management through ICT, 61–70.

started around the same time, the focus of VRC was exclusively on vocational rehabilitation.

With this historical background let us first examine what is meant by vocational rehabilitation. As accepted by the International Labour Organization through its recommendation in 1959 and later in 1983, the term 'vocational rehabilitation' means that part of the continuous and coordinated process of rehabilitation which involves the provision of vocational services, such as vocational guidance, vocational training and selective placement, designed to enable a disabled person to secure, progress and retain suitable employment.

The three important components of the vocational rehabilitation process are vocational guidance, vocational training and selective placement in wage paid employment, self-employment or supported employment services.

Any subject dealing with service to population has to be examined with reference to the prevalent demographic profile of the group in the country. The prevalence of disability has been reported differently by different organizations and NGOs. However, the only authentic data used for the purpose is the Census 2001 and NSSO (National Sample Survey Organization) 2002 data. Both the organizations place the prevalence at 1.98 to 2.10 per cent of the total population. The distribution of prevalence is more or less the same in rural and urban areas, but the number of persons living in rural areas is far greater than in urban areas. When it comes to vocational rehabilitation, the status of education and vocational training among the PWD is to be factorized for planning the services.

As per the NSSO illiteracy among the PWD is as high as 54.70 per cent. Only 9.20 per cent have studied upto the Secondary School level. Even here the ratios are different for different disability groups (Table 8.1).

Table 8.1. Literacy levels of different categories of PWD, NSSO study (per 1000).

Disability	Illiterate	Primary	Functionally illiterate	Middle	Secondary & above
Mental retardation	866	106	97.20%	24	3
Mental illness	591	215	80.60%	102	89
Blindness	773	141	91.40%	45	41
Low vision	738	172	91.00%	45	43
Hearing	646	231	87.70%	70	50
Speech	670	235	90.50%	57	38
Locomotor	447	292	73.90%	136	124
Any Disability	547	254	80.10%	106	92

We are aware that for all practical purposes those with primary education are as good as illiterate. If this is taken into account the percentage of really educated who could be given any statutory level training is only 19.80 per cent.

The Muthukumaran Committee on education in the educationally forward state of Tamil Nadu found that even in 2007 there was a drop-out rate of about 70 per cent *among non-disabled* at the Secondary School level. I leave it to imagination the literacy rates of the disabled in Tamil Nadu. If we look at the details, we find that the drop-out rate at the +2 level was more than 81 per cent ten years ago. We are now dealing with this group that dropped out of school about ten years ago.

The data given in Table 8.2 is confirmed by our own experience at the VRCs. Normally those registering at any institution for employment opportunities are expected to have some education. On this parameter alone, one would find that over 50 per cent approaching employment-oriented institutions such as VRCs and employment exchanges have had education less than 10th Standard. It was found to be 88.60 per cent in the rural camps covered by the VRCs.

A comparative study of those admitted at VRCs, employment exchanges and at camps in rural areas is shown in Table 8.3.

In a more recent study by the author (November 2007) at the District Disability Rehabilitation Centres (DDRC) in the country, see Table 8.4 selected

Table 8.2. School drop-out rates in Tamil Nadu.

Level	Now (2007) (in %)	10 years ago (in %)
Primary School (5th Std)	2.7	15.85
Middle School (8th Std)	5.22	32.54
High School	42.55	63.87
+ 2	69.45	81.4

Table 8.3. Comparative study of educational and vocational skills of PWD.

Education	VRC (in %)	Emp. Exch. (in %)	Camps in rural areas (in %)
Illiterate	9.9	0.4	40.3
< 8th Std	25.7	12.7	42.1
Less than 8th	35.6	13.1	82.4
8th to 10th	14.5	21.8	6.2
SSC +	49.9	65.1	11.4
Vocational Skills	20.8	30.2	3.2

According to the NSSO report only 1.90 per cent possess vocational skills.

Table 8.4.   Educational status of beneficiaries of DDRCs.

Education	Number	% age
Illiterate	64	32.50
Literate	4	2.10
<8	81	41.10
Secondary	22	11.20
Sr. Secondary	15	7.60
Graduate	6	3.00
Post-graduate	5	2.50
TOTAL	197	100%

on a stratified sample basis, it was seen that 75.70 per cent PWD have studied only up to the 8th Standard.

Against this background of high illiteracy let us examine the facilities for delivery of skill training services that can be accessed by the PWD. There is not a single institution for providing vocational and other employable skills to the PWD in the country at any statutory level. A few educational institutions run pre-vocational services, but mostly focusing on non-employable or uneconomical trades such as candle making, *newar* making, handloom, hand composing, chalk making, embroidery, and so on.

The PWD are thus forced to access regular vocational training institutions such as ITIs and polytechnics. The 3 per cent reservation accorded to them has not borne much fruit for several reasons, including lack of awareness, lack of educated PWD, resistance from institutions to those other than the locomotor disabled.

Tables 8.5 and 8.6 give a picture of seats available for the disabled in technical institutions.

Even though over 31,000 seats are available for the PWD, the actual utilization is less than 8,000. The main reasons for this are:

1. Lack of awareness about technical education facilities.

Table 8.5.   Seats in ITIs in India as on 11 January 2008.

Region	Govt	Seats	Pvt	Seats	Total ITIs	Total Seats	For pwd @ 3%
North	633	102,425	341	28,416	974	130,841	3,925
South	384	87,999	2,166	232,722	2,550	320,721	9,622
NE & East	167	39,884	221	25,624	388	65,508	1,965
West	712	169,680	490	55,500	1,202	225,260	6,758
Grand Total	1,896	399,988	3,218	342,342	5,114	742,330	22,270

Table 8.6. Polytechnics and other diploma-level courses.

Particulars	Insts	Intake	Available @ 3%
Engineering	1,244	265,416	7,962
Pharmacy	415	24,029	721
Hotel Mgmt	63	4,020	121
Architecture	25	905	27
TOTAL	1,747	294,370	8,831

2. Lack of suitable qualified persons from a few disability groups, especially the hearing impaired.
3. Not all operations and work tasks involved in the trade can be performed by the PWD, especially the visually impaired.
4. Only persons with locomotor disabilities are admitted to these institutions.
5. There is a resistance from institutions about a few disability groups–mainly the hearing impaired–because of lack of knowledge about how to deal with them.

Studies at VRCs, NSSO and other institutions in the country have found that hardly 3 per cent approaching institutions have any vocational skill.

Whether in the field of ICT or otherwise, there is a need for innovation and the imperative of reaching the rural areas. The author had opportunities to take vocational training services to the rural areas through community-based vocational training. It was done by identifying the employment market in the concerned rural/semi-urban area, identifying the skills required and designing the syllabus with weekly and monthly skill targets, training the PWD and helping them get suitable wage earning avenues in the community. The author has been carrying out these programmes since 1988 with success and least sociological disturbance. Group employment, self-help groups and cooperatives have been formed in different parts of the country, as per market demands and needs of the PWD.

Vocational rehabilitation is achieved mainly through the Ministry of Labour, through the Vocational Rehabilitation Centres for the Handicapped. There are at present 20 such centres spread in 19 states/UTs. The major objective of these centres is to identify the residual abilities and help the PWD choose the best feasible vocational rehabilitation avenue suited to their needs.

This is achieved through a systematic evaluation of medical, psychological and vocational abilities on which the career could be built.

The focus is on:

- Functional ability despite impairment
- Capacity to withstand the physical, mental and emotional strain associated with the occupation
- Identify areas with the possibilities to enhance vocational potential
- Identify the vocational needs of the PWD
- Assess skills
- Assess potential for retraining for re-employment
- Assess work environment modification to suit the individual

After the assessment action is taken for:

- Counseling the PWD on employment market
- Parent counseling
- Behaviour modification, to suit employment
- Provide employable skills in addition to the vocational skills
- Adjustment training in operations to make the disabled employable for short periods
- Entrepreneurship Development Programme
- Pre recruitment training for competitive examinations.
- Worker education
- Custom-made skill development as required by the employer
- Retraining of persons with recently acquired disabilities
- Organize community-based vocational training programmes
- Placement of PWD in open employment, self-employment or sheltered employment depending on ability and need.

The other objectives of the VRCs are:

- Job development
- Job identification
- Creating community awareness — advocacy
- Assist NGOs in setting up new rehabilitation facilities
- Networking of employers and NGOs
- Utilize community resources

On completion of evaluation the VRCs place the PWD in suitable jobs in open employment or self-employment or help in getting supported employment services.

The following services are available in open employment:

- The 3 per cent reservation under the PWD Act is applicable only in public sector employment.
- This has been in operation since 1977 and made statutory in 1995.
- Several posts have been identified by the government where PWD can be employed
- Group A: 308
- Group B: 140
- Group A and B: 324
- Group C: 833
- Group D: 95

Only VH (Visually Handicapped)

- Group A and B: 159
- Group C: 149
- Group D: 21

In addition to this there are facilities for starting self-employment ventures through different schemes, such as banking and financial institutions of the different state governments (industrial development corporations, small industries corporations, and so on). A National Handicapped Development Finance Corporation to micro-finance ventures started by the PWD has been set up by the Government of India. It provides loans to the disabled and parents of persons with mental retardation, autism, cerebral palsy and multiple disabilities.

The Government of India also provides assistance to set up sheltered workshops and different NGOs and other vocational rehabilitation institutions are assisting the disabled to form self-help groups, cooperatives, and other group employment avenues.

It should be borne in mind that in India the organized sector consists of only 7 per cent of the work force; 93 per cent of the work force is in the unorganized sector where, according to the National Commission on Unorganized Sector constituted by the Government of India, 79 per cent of the 395 million get wages of less than Rs. 20 per day. The condition of the PWD cannot be any different.

Our focus therefore has to be on providing vocational skills and creating opportunities for supported employment services, to make the PWD as independent as possible.

- Train the PWD to function as a cog in the supply chain for the new business initiatives undertaken by large companies such as Reliance/Bharti/Aditya Birla Group/Spencer's, and so on.
- In the services sector, servicing MNCs and exporting software; tertiary activities related to hardware.
- With the expanding networking and increasing density of telephony in the rural and urban areas, there is an opportunity in the field of service and repair of equipment.
- Train PWD to suit low-technology jobs where education is not a constraint such as repair of mobile phones, selling SIM cards. The village *Chaupal* can absorb a few.
- Train in networking–hardware and software. Present hardware, especially in computers, is replacement and installation which may not require higher education.
- Train PWD to access telework situations where face-to-face contacts are minimal would suit those with severe disabilities.
- By eliminating obstacles to communication the disability may give rise to broader employment avenues and assimilation of the PWD.
- Persons with severe disabilities could be helped to network with others in the community, country and the world for companionship and wage earning. Remember the unorganized sector, which constitutes the major chunk of work force in India, gets less than Rs. 20 per day. If we can help the PWD get at least this amount, we are doing a service
- The author had introduced community-based vocational training in rural areas suited to the local employment markets as early as 1988. Hundreds of such training programmes have been carried out in all sorts of occupations–electrical, mechanical, agricultural, floriculture, mushroom culture, garments, screen printing, and so on for the less educated rural PWD. This has resulted in helping

more than 80 per cent of the trained to earn within three to six months.

- The DGET (Director General of Employment and Training) has come out with modules of short duration for the less literate. Hundred modules have been identified in each zone of the country. Training could be provided in these operations for suitable employment.
- The CCEA (Council for the Curriculum, Examinations and Assessment) has now approved a special scheme where private employers in the organized sector could be compensated. Even though this would only be nominal to the extent of payment of employer's contribution towards Provident Fund and Employees State Insurance for the workers, it is a small incentive.

We have to remember that:

- *The rehabilitation programme is not designed to cope with emergencies; its goal is to assist persons with disabilities to attain the best possible vocational adjustment, not the most expedient, nor the most convenient.*
- Vocational rehabilitation is not a quickfix solution, where persons with disabilities get rehabilitated as soon as they arrive.
- *Jobs are not readily available–jobs are not made for the disabled, and the persons with disabilities him/herself is not fully ready for placement. In a majority of cases of PWD approaching institutions almost 90 per cent do not possess any vocational or employable skill.*
- We have to take a proactive role in creating suitable facilities for evaluating the individual and placement.

All jobs are suitable and no job is suitable.

- A few components of each job can be performed
- Analyse each job
- Identify operations
- List out simple to complex tasks
- Train the candidates

Vocational and skill training are the first steps in the process of vocational rehabilitation. The next step of placement in wage paid employment, self-employment or other supported employment avenues could be and are being taken up as corollary to vocational training.

- The key to reducing inequality is not charity or philanthropy but innovation and empowerment
- Education is not a constraint–only learning abilities are.
- No employer has a job suitable to the PWD. We have PWD to suit jobs.

9

Parent Societies and the Role of ICT in Rehabilitation Support

Dr. V. K. Gautam[1], Meekankshi G. Trivedi
and Indrani Chakraborty

*104, shri sumati apt., 72, geeta nagar coloney, behind St. Paul School, Indore-452001;
vkgautam16@rediffmail.com*

> *To be born as a disabled child is a wish of God.*
> *To be ignored because of disability by the society is a sin.*
> *To be not cared for and helped by the State is definitely a*
> *crime.*

Introduction

A new born child brings happiness to the family. It is looked upon as one who will keep up the family line and an angel of prosperity. Enquiries about the sex of the child come quickly. If it's a male there's greater happiness; if it's a girl child, parents and family members console themselves saying it's God's wish. However, parents who have an abnormal child rarely ask such questions or respond in like manner. When informed about a child birth, their question is: how is the child? Did she/he cry at the time of birth? Is she/he a normal baby?

Rearing a Disabled Child

It is a well known fact in Asian society that rearing a child requires awareness on the part of parents to help children enter mainstream society. But to rear

Disability Rehabilitation Management through ICT, 71–82.

a disabled child is not just confined to the upbringing of the child but an awareness among parents about the right norms to bring up such a child, which includes awakening society at large. The onus of this lies on the parents. After having broadened their mind, they must now share their knowledge with their family members and the neighbourhood. This could be a small beginning to develop an integral support system.

Evolution of Rehabilitation

The evolution of treatment strategies towards persons with disabilities has passed through four main stages. These stages have been influenced by the human value system, religion, and scientific and technological development, especially IT (see Figure 9.1). Before we go into the history of rehabilitation, we should understand how many people are affected by disabilities. The number of persons with disabilities is given in Table 9.1.

(a) Stage 1: **Society's Impact** — Many denied rights to disabled persons.

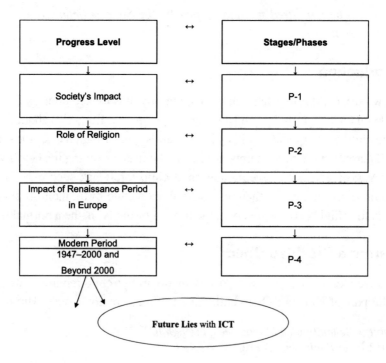

Fig. 9.1 Phases in the History of Rehabilitation.

Table 9.1. Demography of Disabilities (2008).

Type of disability	Urban	Rural	Total
In seeing	27,61,498	78,73,383	1,06,34,881
In speech	3,97,014	12,43,854	16,40,868
In hearing	2,38,906	10,22,816	12,61,722
In movement	14,50,925	46,54,552	61,05,477
Mental	6,70,044	15,93,777	22,63,821
Total	55,18,387	1,63,88,382	2,19,06,769

Table 9.2. Schools for Various Disabilities: 1947 and 2000.

S. No.	Disabilities	No. of schools in undivided India	No. of schools in 2000
1	Visual impairment	32	400
2	Deaf	30	900
3	Locomotor impairment	Nil	700
4	Mentally challenged	03	1,200

(b) Stage 2: **Role of Religion** — Society begins to recognize them.

(c) Stage 3: **Impact of Renaissance in Europe** — Industrial Revolution makes an impact.

(d) Stage 4: **Modern Period** — We can divide this period from 1947 to 2000 and beyond. Until 1947, India lagged behind the Christian missionaries in the area of special education, volunteer organizations (VO) looking after this most neglected area of social work. After Independence, the Union and state governments, with the help of VOs/NGOs, have made quantum progress, as evident in Table 9.2.

Until the mid-950s, the basic philosophy of special education and rehabilitation was to keep the disabled segregated from the community. Thereafter, major changes took place, due to two main factors:

(a) **Involvement of agencies:** Several agencies started experimenting by placing disabled children in regular schools, but the latter were again the main sufferers.

(b) **Union Government's initiative:** The government established Social Employment Exchanges and sought the help of the International Labour Organization in developing vocational training programmes. In 1987, the Social Welfare Ministry announced a 3 per cent reservation in Class III and IV government jobs for the blind, deaf and physically handicapped, but made no similar provision for the mentally challenged. In 1995, the Equal Opportunities, Protection of Rights and Full Participation Act was passed. Another breakthrough was when the government enacted a law for care of persons with mental retardation, cerebral palsy, autism and multiple disabilities through the National Trust Act, 1999.

Disabled/handicapped Persons are Inseparable from Society

The development of disabled persons is a social and psychological need as well as an economic necessity. There are numerous institutions in India engaged in providing social and educational support for rehabilitation. These institutions can be classified as: (a) residential, (b) early intervention centres, (c) daycare centres, (d) special schools, (e) vocational training centres, (f) special clinics, (g) wards in hospitals, (h) common schools, (i) teacher training centres, (j) parents association and cooperative societies of disabled persons (Like Parivaar — National Federation of Parents' Associations, 1995).

Government's Role

The government has paid attention to the education, training, employment and rehabilitation of the disabled. The State provides grant-in-aid schemes but restricted only to education and vocational training and limited to the salary of the trainers and staff of organizations engaged in the field. These institutes depend upon other sources like donations, project grants and sponsorships. The state of Maharashtra In India, which is industrially, economically, socially and culturally developed, has done commendable work in this field. All the four southern states are also doing good work. An increasing amount of efforts are now being undertaken by the central, northern and north-eastern zones, the government as well as Parivaar.

About Parivaar

Its Network

Parivaar is the largest parents'-run association with 210 NGOs associated from all the 35 states and Union Territories. These NGOs work in the field of autism, cerebral palsy, mental retardation and multiple disabilities. Parivaar has conducted a national-level conference every year. Besides this, two Executive Committee (EC) meets, 2–5 regional symposia/workshops/seminars are being conducted on a regular basis every year. In 2007, Parivaar conducted four Regional Parents' Meet at Cochin, Indore, Rohtak and Guwahati, sponsored by National Institute of Mental Health, Secunderabad and two consultative conventions at Lucknow and Shillong sponsored by (Christian Blind Mission) CBM, Germany. In 2010 as well the same number of regional parents' meets were held. Parivaar-NFPA was formed in the year 1995 with a handful NGOs having expanded at a fast pace. Parivaar is a grassroots-level organization with 72 urban, 46 semi-urban and 4 rural parents' associations, often formed by families living Below the Poverty Line (BPL). The target set for 2012 is to cover all the districts of India (See Appendix A).

Its Vision

Creation of awareness within society to accept the intellectually disabled and those with other developmental disabilities as contributing members at par with others and to eventually enable their inclusion in every sphere of life.

Its Mission

Advocacy on behalf of persons with mental retardation, cerebral palsy, autism and multiple disabilities to gain and protect their rights; lobbying to provide more concessions and schemes to improve their quality of life; secure the right for inclusive education to make them part of the mainstream and create an environment for them to live independently to the extent possible in the community.

Its Objectives

- To secure legal rights for persons with mental handicap, cerebral palsy, autism and multiple disabilities and their families.
- To fight for human rights and social justice, against exploitation, abuse and discrimination by implementation of legal rights of persons with these disabilities and by appropriate representation for Parivaar-NFPA in policymaking bodies at the central, state and at local levels.

- To promote the interests of persons with mental handicaps, cerebral palsy, autism and multiple disabilities and their families by bringing about cooperation and coordination among organizations at the state/district level.
- To make an endeavour for bringing about the inclusion of the disabled in all fields of life with a focus on inclusive education.
- To create a common bond of understanding and action among parents, families, professionals and others concerned with the problems of the above disabilities, throughout India and the world.
- To protect the rights of persons with disabilities and their parents, families, guardians and enable them to live a life of dignity and enjoy full citizenship rights.
- To make an endeavour to bring all parents under the common umbrella of Parivaar-NFPA by forming parents' associations and training them to function as pressure groups to achieve the above objectives.
- To create public awareness through print and electronic media.

Its Organizational Structure

To attain its objectives, Parivaar is equipped to execute any such programme provided financial help is received. Accordingly, Parivaar's organization has been divided into six zones, each headed by a vice-president and a zonal secretary. Its organization structure with allied functionaries is given in Figure 9.2.

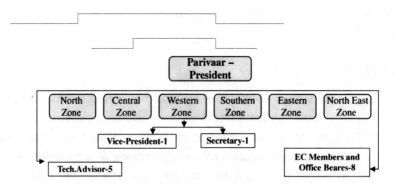

Fig. 9.2 Organization Structure of Parivaar (Maximum of 41 Members in the Executive Council).

Need to Penetrate the Interior

Most NGOs are eligible for tax exemption under section 80 G of the Income Tax Act, 1861, but the most affected population lies in the rural areas which are Below Poverty Line. It has been experienced that Parivaar lacks in awareness, development of self-help groups (SHG) and leadership (especially second-rung leadership), preventing the expansion in its scope and functions and thereby allowing penetration deep within the country's interior.

Disability and Rehabilitation

Some Definitions

The terms 'disability' and 'handicapped' are used interchangeably. 'Disability' refers to inference in functioning and is therefore more apt to use, because it indicates that an individual is unable to do something. 'Handicap' refers to a medical model: disease–impairment–disability–handicapped intellectually or developmentally. As per the World Health Organization, impairment refers to any loss or abnormality of psychological, physical or anatomical structure or function. Impairment may involve difficulty in walking or speaking needs. A disability becomes a handicap when it interferes with doing what is expected at a particular time in one's life. Impairment represents a missing or defective body part. The Persons Disabled Act, 1995 defines disabled as, 'A person suffering from not less than 40 per cent of any disability as certified by a Medical Authority.'

Importance of Terms

We will restrict ourselves to limited disabilities for a better understanding. It is imperative that we understand a few terms like impairment, disability, handicapped, mental retardation, cerebral palsy, autism, multiple disabilities and Alzheimer's Disease (perennial dementia characterized by confusion, memory failure, disorientation, restlessness, speech disturbance, and inability to carry out purposeful movements. It is generally experienced by the intellectually disabled after 30 years of age, while a normal human may be affected between 66 and 72 years of age.

Rehabilitation of the Disabled: Implications involved

Rehabilitation has certain implications related to an individual or an institution:

(a) **Individual-related implications:** They include difficulties and limitations of the disabled.

(b) **Institution-related implications:** To provide prevention, and give respite.

Role of ICT

IT in Rehabilitation of the Mentally Challenged

In this essay, efforts have been made to show how ICT can help the disabled rehabilitate themselves. Considering the extent of IT facilities available in the country, every disabled may not be able to access them, but training centres can make use of the models and recommendations given in the following paragraphs.

Computer Based Training (CBT) and Rehabilitation

CBT includes use of computer for routine work, use of multimedia, credit card or any e-card, digital video learning, full-motion video on CD-ROM for training, CBT for testing, theatre-based training and video-conferencing.

Preparing for Premature Old-age

After 32 years of age, mentally challenged persons start becoming non-functional or start looking 20 years ahead of their normal counterparts due to Alzheimer's Disease, Down Syndrome and other such mentally challenged. Thus, they start ageing before time. Parents can download literature from the Internet and help keep children involved in as many activities as they can, especially social rehabilitation.

Parents as Communicators

There are numerous case studies in India, of 'A parent as a Communicator' wherein the mother has aided in the development of the child — physically as well as speech-wise handicapped — in such a way that the latter is able to appear on stage, due to exposure and confidence-building exercises.

Diagnostic Care

IT enables the provision of medical help. If need be, help can be sought from specialists abroad provided we are in a position to communicate via the Internet.

Cyber Cafés

Some cities have cyber cafes every 200 metres, which a PWD (person with disability) person can maintain for a few hours. These cafés can be installed at vantage points like railway stations and public places near hostels, hospitals and institutions. However, a technically aware person has to be present as an observer/monitor.

IT Centres

Here the job involves taking care of the security of the machinery and hygiene. At university levels, these jobs can be part of maintaining IN/OUT registers, checking the specific gravity of electrolyte and topping up with distilled water, checking voltage of batteries used for backup, especially for servers.

Video Parlours

Video games are very useful for improving the reflexes. Mentally Challenged people with an IQ score of 50 can easily help in attending to customers, issuing discs, keeping an account of time, and so on, provided he/she has an aptitude for this kind of job.

Disability Rehabilitation–PCO

The government has several provisions to give PCOs to the disabled on priority, provided they can read the bill and count currency. They can also be employed to operate photocopiers or any such safe machines.

Media-Home Platform

It can be used to nurture/develop talents. The content is amusing and presentation-attractive, especially animated figures. It helps to change habits, as 'learning by observation'. (See Figure 9.3). It is connected to the Internet

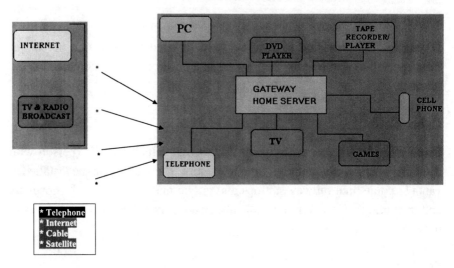

Fig. 9.3   MHP Networking.

and gadgets like the computer, Tale recorder, DVD, TV. Initially people are trained by a trainer and subsequently, a PWD with even an IQ level of 40 and above can operate and improve learning skills.

E-Manthan

There are numerous centres like e-manthan for typing work. E-manthan pay Rs. 5 to Rs. 10 per page. Many more such companies are coming up on the lines of medical transcription. Getting typing work done, as is carried out for operations of date entry during elections to update records, is one such example of this form of work.

Booking Window

Any hearing impaired or MC (IQ above 70) person can man a booking window for fixed types of jobs or where written/non-verbal rather than verbal communication is required. Today, all railway stations have been computerized and entries are limited as far as middle size stations are concerned. In such reservation windows, PWD with 55 and above IQ can assist the customer to give request information about the seats availability and timings as information is pre-programmed.

Inhome and In-lab Rehabilitation

With basic training and under supervision, a MC (with 70 plus IQ) can help educationists or scientists save time by operating computers by way of booting, connecting, opening of Internet, initial surfing, and so on. They can help in taking prints, scanning photographs, preparing slides, and so on, but 'aptitude and skill are musts'. Children have to be made computer-savvy, along with many mothers preparing, especially female children, for household activities.

Voice Recognition and Touch Pads

There are software(s) for voice recognition and touch-pads. Touch-pads are suitable for the visually handicapped; voice-recognition software can understand the voice and write or give desired command. This software can sensitize the machine and work perfectly.

Technocrats Advice on use of IT for the Mentally Challenged

An opinion survey was carried out in the University Campus of Indore in 2003 and 2007, where 180 students and professionals and 34 faculties of International Institute of Professional Studies (IIPS) and School of Computer Sciences (SCS) of (DA University) participated. Though none was directly or indirectly affected but they gave their viewpoints frankly and freely. Their recommendations were:

(a) Development of visual software to make their words understood, to keep records and track the improvements made by such people.
(b) The computer can sharpen brain by special application and creating interest in machines.
(c) IT companies can provide small jobs like advertising and installing kiosks at vantage points.
(d) IT can design tailor-made software to train people for rehabilitation.
(e) Blinds and embedded keyboards can be designed and provided for the visually handicapped.
(f) They can be employed in cyber cafes for selective jobs.
(g) Mental retardation can be improved by improving their reflexes.

(h) They can be trained with the help of Powerpoint and made to prepare Powerpoint slides.

(i) They can surf, search and store pictures. This can provide a lot of information, that too speedily.

(j) They can be stationed at ATMs as observers and to assist new customers using the machine for the first time. (In any case one guard is always there.)

(k) They can write, copy, rewrite CDs and do similar jobs involving limited and fixed operations.

(l) Recommended jobs at call centres.

(m) Asked to search on the Internet for such opportunities abroad.

(n) Data-entry

(o) Co-opted into family business, irrespective of their profession.

Conclusion

Growing and developing are ongoing processes and the strength of a nation lies in the weakest link of society. Hence, it is the duty, not only of the government but every one of us to create an environment to prepare the diabled people to face the challenges of disability and to make them self-reliant. Notwithstanding the fact that networking is a key to success, the able segment of the society must use the advancement in ICT and its tools for improving the quality of life of the disabled.

Part III

ICT Initiatives for Rehabilitation of Differently Abled Persons

10

Empowerment is About People Taking Control Over their Lives, Becoming Conscious of their Own Situation and Position, Setting their Own Agenda, Creating Self-Confidence, Solving Problems and Developing Self-Reliance

Dr. B.S. Bhandari

References

* Parivaar [NFPA (National Federation of Parents Associations)], and re-designated in 2010 as NCPO-National Confederation of Parents' Organizations, New Delhi) is the largest association run by parents with 210 associate NGOs, working for people with autism, cerebral palsy, mental retardation and multiple disabilities. It is a grassroots' level organization with 104 urban, 56 semi-urban and 50 rural parents' associations, most often formed by families living below the poverty line.

The disabled need to be a part of society and be treated equally, enjoying the benefits of life and opportunities of development. NGOs may be organized by parents and siblings to secure legal rights for such handicapped persons with the help of government and community through networking. This essay outlines the importance of networking from the viewpoint of parents' associations to cope with and access the services rendered for the betterment of the quality of life of mentally challenged persons. NGOs can attain their objectives

Disability Rehabilitation Management through ICT, 85–87.
© 2012 *River Publishers. All rights reserved.*

successfully and effectively only when they are well informed about the needs of mentally challenged children and their families interested in taking care of them. These needs may be fulfilled by the parents, community, specialists, physicians, institutions, business houses and government. For this purpose, networking and development of structural framework are significant. This organizational structure may not work effectively unless appropriate supportive action and flow of support is mobilized. This can be best achieved when the system of information and communication technology is properly introduced, as digital technology and modern methods of communication have made this very effective. This essay also points out different aspects of ICT which provide direct support to NGOs in rehabilitations activities for the handicapped.

*Lt. Col. (Dr.) V.K. Gautam, Parent, General Secretary, Parivaar, and Group Director, Modern Group of Institutes, Indore;

**Meekankshi G. Trivedi, Sibling and Lecturer, J.G. College of Education, Ahmedabad; e-mail: ID-meenakshi98@rediffmail.com

***Indrani Chakraborty, Parent, Special Educator, and Activity Professional at Samarth Center, Guwahati; e-mail: indranic@sify.com

References

[1] H.P.S. Ahluwalia, *Training Manual on Disability Management and Mainstreaming of Persons with Disability for IAS Officers*, New Delhi: Rehabilitation Council of India, 2001.

[2] Background Papers (9-10 October 2003), released by Government of India, (MSJE).

[3] C. I Chakraborty and M. Trivedi, A Case Study-Intellectual Disability: A Challenge to a Mother and a Family, published in "Opportunities and Challenges in the Global Business", EXCEL BOOKS, New Delhi, 2009, pp. 655–7.

[4] Henry Enns, *The Role of Organizations of Disabled People*, www.indepentliving.org.

[5] Dr. A. Mervyn Fox, *An Introduction to Neuro-Developmental Disorders of Children*, 1st Ed., New Delhi: The National Trust, 2003.

[6] 'Mainstreaming Persons with Disability: Role of Volunteering and Media', produced by Sanjeevan Enterprises, New Delhi, 2001.

[7] M.D. Nichternsol, *Helping the Retarded Child*, New York: Grosset & Dunlop Publishers.

[8] Dr. A.K. Ramani, Jaydeep Tripathi, and Gagn Bindra, 'Human Resource Planning for IT Application in the State of MP', unpublished paper, September 2002.

[9] Romayne Smith, *Children with Mental Retardation: A Parents' Guide*, Bethesda, MD: Woodbine House, 1993, www.aamr.org.

[10] Dave Zieliski, 'The Complete Training Library', *Using Technology-Delivered Learning*, Vol. 1.

[11] Australia Rehabilitation and Assertive Technology Association (ARATA) [www.e-bility.com / arate OR www.e-bility.com]

[12] www.tribuneindia.com-by Champion, Dr. E.M. Johnson of RCI North Zone, 2003.
[13] www.cbrnetwork.org/cbrservlets

Appendix A

Table A.1. Demography of India with Regard to Population, Area and State-Wise Number of NGOs Affiliated to Parivaar.

Sr. No.	States	Population	Area (Sq.km)	District	NGOs affiliated to Parivaar
01	Andaman & Nicobar Islands (UT)	3,56,152	8,249	2	Nil
02	Andhra Pradesh	76,210,007	2,75,069	23	23
03	Arunachal Pradesh	1,097,968	83,743	16	NIL
04	Assam	26,655,528	78,438	23	03
05	Bihar	82,998,509	94,163	37	01
06	Chandigarh (UT)	900,635	114	1	02
07	Chhattisgarh	20,833,803	135,191	16	02
08	Dadra & Nagar Haveli (UT)	220,490	491	1	Nil
09	Daman & Diu (UT)	158,204	112	2	Nil
10	Delhi	13,850,507	1,483	9	08
11	Goa	1,347,668	3,702	2	01
12	Gujarat	50,671,017	196,022	25	05
13	Haryana	21,144,564	44,212	20	01
14	Himachal Pradesh	6,077,900	55,673	12	02
15	Jammu & Kashmir	10,143,700	101,387	14	01
16	Jharkhand	26,945,829	79,714	22	03
17	Karnataka	52,850,562	191,791	27	20
18	Kerala	31,641,374	38,863	14	27
19	Lakshadweep (UT)	60,650	32	1	Nil
20	Madhya Pradesh	60,348,023	308,245	48	11
21	Maharashtra	96,878,627	307,713	35	37
22	Manipur	2,166,788	22,327	9	01
23	Meghalaya	2,318,822	22,429	7	01
24	Mizoram	888,573	21,081	8	02
25	Nagaland	1,990,036	16,579	8	NIL
26	Orissa	36,804,660	155,707	30	05
27	Pondicherry	974,345	480	4	Nil
28	Punjab	24,358,999	50,362	17	05
29	Rajasthan	56,507,188	342,239	12	03
30	Sikkim	540,851	7,096	4	01
31	Tamil Nadu	62,405,679	130,058	31	23
32	Tripura	3,199,203	10,486	4	Nil
33	Uttaranchal	8,489,349	53,483	13	03
34	Uttar Pradesh	166,197,921	240,928	70	02
35	West Bengal	80,176,197	88,752	19	17
	Total (approximate)			586	210

A few NGOs in the pipeline for affiliation or under consideration for want of documents (It include parents as well associated organizations and NGO from J & K became member in 2010).

11

IT-Medical Rehabilitation: A Tripura Case Study

Dr. Ashesh Ray Chaudhury

The radio telescope allowed man to explore areas of the universe that the old optical telescope couldn't reach, going thousands of light years further than the latter. The telescope is perhaps a good metaphor for information technology, which has expanded the service provision of medical rehabilitation from the institute-bound expertise to the farthest rural corner.

Among the important shortcomings in medical rehabilitation services in India, the first is the incomplete and poor database. Service is mostly institute-bound, which, by definition cannot be available to too many. The disabled remain uninformed of the disability management and, even if treatment starts, it cannot continue either out of geographical or communication constraints.

It was to overcome the constraints of a low cost system that it was planned to use IT. The author put the system to work since 2005 August.

It is an efficient data storing and readily available retrieval method on the one hand and an easily accessible method for the beneficiaries on the other. The former, which is the controlling or operator's end, may be referred to as the 'doctor's end' and the other end, where the beneficiaries use it, may be called the 'patient's end'. The connecting system lies in the middle.

Disability Rehabilitation Management through ICT, 89–92.
© 2012 *River Publishers. All rights reserved.*

The ***Doctor's end*** is made of the following:

1. A computer, preferably a laptop with an Internet connection; the computer is to run the program developed for this purpose.
2. A mobile phone. One advanced phone with GPRS facility and mini excel running ability (palm top) will give greater mobility to the doctors and easy reach for the patient.
3. Recently the software program is modified to be hoisted in one server so that from any place all data can be accessed by anyone who has the permission to do so and has internet access. It may be possible with a mobile device with fast net connectivity. This is going to be operational soon.

The ***patient's end*** can work with any phone or computer, if available.

The usual telephone system, either the land line or a mobile phone (GSM or CDMA), is the connecting system. Internet connectivity is required for the doctor and the patient. Although it is optional for the latter, it allows the patient to receive the doctor's prescription and instructions in detail. Otherwise, this can also be done over the phone or instructions can also be sent as 'SMS' to the mobile phone.

The unique software developed by the doctor has the following key features:

1. Needs only the easily available MS operating system.
2. It records all the general details of the patient like name, address, body mass index, and so on.
3. Clinical details can also be recorded in detail, with a drop-down menu, without any typing.
4. Prescription can be composed easily, without any typing except the diagnosis. Prescription may be prepared just adding the templates made out beforehand according to the need and wish of the pre-scribing doctor or adding from the table of medicines, orthosis or exercises etc made out as required by the doctor. It takes hardly a minute or two to print the prescription for the patient. This can be done when the patient is physically present before the doctor.
5. Patients' records are stored in classified forms.

6. The stored data can be retrieved in many ways — by the unique id or name or the diagnosis or the date of recording.
7. For their follow-ups, patients can consult the doctor over the phone or the Internet, even from the farthest areas. They are not required to visit the doctor physically.
8. Although the data is recorded in Microsoft Access, it can be transferred to an Excel sheet and be carried in the PDA mobile phone to answer the patient's call from any place at any hour. This eliminates the possibility of any error because the doctor does not have to respond from memory.
9. The software program can also send reminders to patients to go for the maintenance and repair of the orthosis — prosthesis or follow-up.

Operation: The first record should be created when the patient is physically present to meet the doctor and all the details are entered into the latter's database. Patients can later ask for advice over the phone, by e-mail, or through the website. The doctor can respond by calling the patient over phone or simply with a 'SMS' from the mobile phone after checking the necessary records in his/her database.

Advantages: Patients can reach the specialist at any time, even if they live far from the institute and the expert. Being so easily accessible, however, patient's call will not infringe on the privacy of the expert or the doctors at the centre. Patient can maintain a life long contact well and his/her development be monitored by one expert working in different centres. Patients can be easily called to any remote place to counsel or assess even from a far off centre. It can be done in much better way if video conferencing system can be added with it; of course increasing the cost of it. With the gradual inclusion of newer patients of one locality an epidemiological picture can be easily drawn from the database.

Otherwise the system is of low cost and can be easily operationalised without complication and elaborate installation. The operator also does not require any rigorous training.

Disadvantage: The system already in use does not have any video conferencing mechanism to keep it simple and low cost, which will, in fact, make it more useful to monitor from a distance. For the initial assessment, patients have to be physically present before the doctor.

Sending prescriptions and instructions to a remote village often seems to be a problem as many villages are not well connected by land or mobile phones. The other problem is that patients have to take help from someone who knows English. That said, more villages are now gradually being brought within the folds of the telephone network. And, as far as the problem of language is concerned, a partial solution is to start the service in regional languages.

Conclusion: The system is being tried in Tripura, where, due to the hilly terrain and geographical location, physical communication isn't easy. The Internet connects most block offices, called CIC. Unfortunately, however, people haven't taken to it yet. If the use of Internet were to become more popular, this low cost IT-based medical rehabilitation service is going to be an obligatory mandate for the future.

12

Tele-Ophthalmology and Preventable Childhood Blindness: The *KIDROP* Experience of Narayana Nethralaya

co-authored by Sherine Braganza, Ravindra Battu, Rohit Shetty, K Bhujang Shetty and Clare Gilbert

Retinopathy of Prematurity (ROP) is a potentially blinding disease that affects premature infants in both eyes. In its advanced stages, the untreated disease can result in permanent and complete blindness. ROP is the leading cause of childhood and infant blindness in the developed world. In India, the prevalence of any stage ROP in high-risk babies has been reported to be around 47 per cent [1]. It is believed that India, along with other developing countries is experiencing the 'third epidemic' of this disease and it is not before long that the absolute number of ROP blind infants in our country may reach those seen in Latin American and other Asian countries [2].

The magnitude of the problem can be gauged by a closer look at the government's census report. In 2008, roughly 27 million live births were recorded in India. Of these roughly 8.8% were estimated to be born below 2000 grams and are at risk of developing ROP [3]. The number of premature babies born in India are often underestimated especially in rural areas where mothers are unsure of dates and do not undergo antenatal ultrasound scans. Yet, it is

Disability Rehabilitation Management through ICT, 93–110.
© 2012 *River Publishers. All rights reserved.*

believed that close to 2 million infants each year need ROP screening (this does not consider infant mortality rate) Approximately 47–54% of these babies may develop 'some disease' and about 10–15 per cent of these children have the potential of going blind if untreated [4].

Introduction

ROP has become a public health problem and an important cause of childhood blindness in the past decade. A high birth rate, increasing neonatal survival and better neonatal care has increased the potential 'at-risk' babies. Furthermore, varying levels of neonatal and eye care, inadequate trained manpower, inadequate treatment resources (ROP is treated with laser photocoagulation using an indirect ophthalmoscopy delivery system that needs considerable expertise) and an inadequate coverage of the ROP screening programme in India thus far has worsened the situation [5]. Ironically, relatively heavier babies (i.e. between 1,500 and 2,000 grams) considered at 'no' or 'low' risk in the developed world, have been shown to develop severe ROP sufficient to cause blindness in our country [6]. This increases the number of babies requiring screening in our setting.

ROP screening allows a very small window of time. The disease may start within three weeks after life in babies born <2,000 grams (and or < 35 weeks of gestation) and within the next 6–8 weeks may progress to complete retinal detachment (i.e. stage 5). Appropriate screening and timely treatment using the Early Treatment for Retinopathy of Prematurity (ETROP) guidelines [7] will result in excellent outcome (approximately 90 % anatomical success) which has been similarly reported from studies in India too [8]. The problem lies in the fact that there is no way to predict which premature baby will develop ROP and which one will escape it. Hence it is recommended that all babies born <2,000 grams [9] be screened by a trained ophthalmologist (who has ROP training) at least once 3–4 weeks after birth and thereafter based on the clinical findings at each visit.

With a less than 400 trained retinal surgeons in India (Vitreo-Retinal Society of India membership) and less than 15 centers capable of comprehensive ROP screening and management services all over India, the challenge lies in using these limited resources to provide screening (and treatment) to the under-served areas of our country [10].

In April 2008 Narayana Nethralaya, Bangalore, initiated a Tele-ROP service using the principles of ICT. This report presents a summary of 28 months experience of this project namely — KIDROP, (Karnataka Internet Assisted Diagnosis of Retinopathy of Prematurity).

Method

The strategy of the KIDROP model can been summarized by the *'Triple T' approach*. The first 'T' stands for *Tele-ROP*; the second for *Training* of peripherally located ophthalmologists to upgrade their skills to allow appropriate screening and treatment of ROP; the third refers to *Talking* to peripheral paediatricians, neonatologists and gynaecologists who are capable of counselling the mothers and family members for ROP screening.

The pilot project was initiated around a radius of 300 km from Bangalore city in Zone 1 as depicted in Figure 12.1. What began as a pilot of 6 neonatal care centres, today has expanded to include 23 centres (October 2010). Centres are visited on a fixed routine once a week or fortnight (in some cases bi-weekly), by the KIDROP team. The team comprises of a trained technician (trained using extensive in-house and on-site sessions with validation), a manager, a driver and an (in some cases) an ophthalmologist. At the centre of the technology is the RETCAM SHUTTLE (Clarity MSI, Pleasanton, California, USA). This is a portable, wide-field, pediatric digital imaging camera capable of imaging upto 130 degrees of the retina of neonates. A portable laser-indirect ophthalmoscope is also carried required to deliver treatment. The equipment and the team members use an SUV (Toyota Qualis) for transport shown in Figure 12.2. The team travels roughly 1500 kms every week to complete one cycle.

Once the team arrives at the NICU (neonatal intensive care units) the Retcam Shuttle is transported inside the NICU to the bedside of each sick premature infant. In addition, neonates who have been discharged or belong to other NICU's in the geographical vicinity also collect at the centre and undergo imaging as shown in Figure 12.3. The examination and imaging sessions are carried out at the bedside of the infant, most of whom continued to be housed in incubators and the mothers are counselled about the disease after showing them the images of their own babies as shown in Figure 12.4. This improves compliance and follow-up.

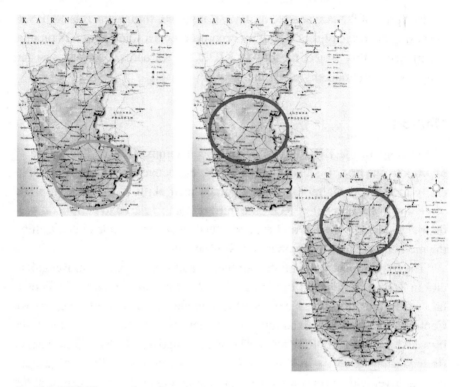

Fig. 12.1 KIDROP – area of coverage.
The blue circle depicts the area of coverage in the pilot project. The red circles depict the zones that are included in the public private partnership with the National Rural Health Mission, Govt. of Karnataka.

Technique

The RETCAM allows imaging of the anterior and posterior segments and the angle of the eye with the help of a variety of lenses. For the purpose of this study, the anterior segment is imaged without any lens attached (i.e. the camera held away from the eye and focus obtained using the foot switch) and the retinal images are obtained using the 130 degree (ROP) lens.

The infant is wrapped in a 'mummified or swaddled' position, under standard aseptic precautions and placed on a trolley with an adjustable height. In cases where the baby cannot be removed from the incubator, the imaging session is carried out inside the incubator with an extension of the infant tray.

Fig. 12.2 Retcam Shuttle transported in a SUV.
The Retcam shuttle (Clarity MSI, California, USA) is the portable infant camera that is transported between all the participating centres in a SUV.

Fig. 12.3 Mothers and their premature infants awaiting ROP screening.
Mothers from centres close to a participating rural centre accumulate to get their premature infants screened by the KIDROP team at a fixed time every week.

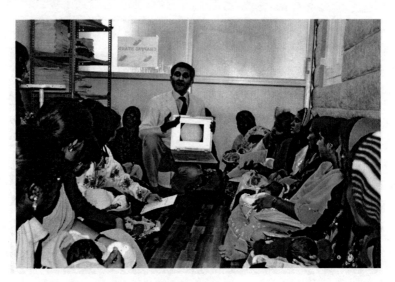

Fig. 12.4 Mothers getting counseled using digital images.
Mothers are shown the images of their infants' retinas and appropriate counseling is provided. Showing them images increases follow-up.

During the entire procedure, the vital parameters are measured in the presence of a neonatal nurse or a neonatologist.

The eyes are prised open with the help of an infant wire speculum. A coupling gel (e.g. Lubic Gel, Neon Lab, Mumbai, India) is placed over the cornea and over the tip of the detachable Retcam lens. The lens tip which is mounted on the portable hand-probe is then carefully lowered over the cornea. The image is brought into focus using the foot switch. The images are saved in a continuous video mode. At the end of the session, the video is reviewed (by the technician) and the relevant stills saved in the database. (Figure 12.5). The whole process takes less than 5 minutes for both eyes.

The technicians capture a set of seven images per eye (more in a quadrant with pathology at his/her discretion), using a modified version of the photo-ROP guidelines [11]. These images include one image of the anterior segment, disc in the centre, macula in the centre, and one each from the nasal, temporal, superior and inferior fields. In these peripheral fields, the ideal image is the one that captures the ora serrata in that quadrant or at least the 'disc-out-of-view' of the opposite extreme border (or just visible) in the image.

The technician saves the images on the laptop connected to the RETCAM, which had an in-built software.

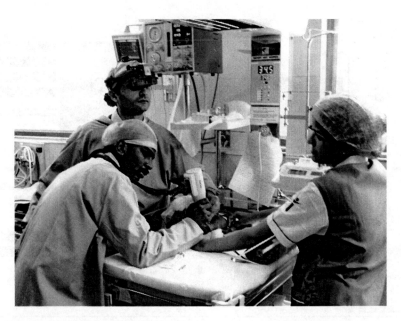

Fig. 12.5 Infant undergoing RETCAM imaging.
The Retcam probe has a wide-field lens mounted on the tip which captures upto 130 degrees of the retina.
Non-Ophthalmologists have been trained to image, capture, store and analyze these images.

Using this software, the technician process the images (using the simple software provided in the RGB modification tool) to highlight various lesions or aspects of the images he had acquired. The software also allows comparison between two eyes or between two imaging sessions. (Figure 12.6)

The technician 'analyse' the images using a three-response triage format: (Figure 12.7)

1) RED — Urgent referral to an ophthalmologist (i.e. inferring that the disease imaged requires treatment or very close follow-up)
2) ORANGE — Needs follow-up (Stage of ROP or immaturity that is not alarming and does not warrant treatment, just yet)
3) GREEN — Can be discharged (Completely mature retina with retinal vessels traced upto the ora serrata in two successive visits):

The first 1007 infants were also examined by one of the authors (AV) using the 'gold standard' of 'binocular indirect ophthalmoscopy, BIO' with peripheral scleral depression and the findings were recorded masked to the findings of

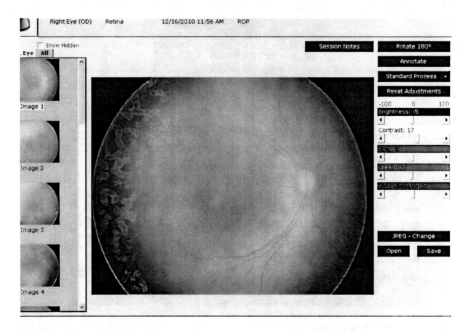

Fig. 12.6 Software of the Retcam.
The software on the Retcam allows easy modification of the images obtained and allows comparison between two sessions which helps the technicians highlight the disease and arrive at a decision.

the technicians. The validation of the technicians was based on this early comparison cohort. Thereafter, the technicians image and record their images independently and these images are uploaded using the methods described below for the remote expert to view and opine on. It is to be noted that the decision to carry out treatment (whenever required) was based on the findings of the ROP expert and not on the technicians recordings in the pilot cohort for medico-legal reasons.

Laser treatment (whenever required) was carried out at the bedside of these infants using the ETROP guidelines, including through the incubator wall in infants who could not be removed from the incubator due to technical reasons [12]. Written informed consent and hospital approval was obtained in each case.

After validation, images are uploaded by the project manager simultaneously onto a specially hosted server backed up in two geographical locations. The upload is performed using a separate laptop (not included with the RETCAM) and a broadband data card for internet access after transfer from the

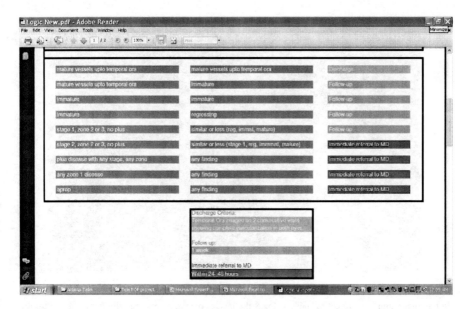

Fig. 12.7 The decision algorithm.
The three colored decision algorithm allows the technician to flag the baby as 'red', 'orange' or 'green' based on the images he or she has obtained allowing triage of the infants seen in the rural peripheries.

Retcam laptop in bmp or jpeg format. (Figure 12.8) The newer machines now have DICOM capabilities and are more robust for transfer and storage.

The ROP expert has access to these uploaded images from any geographical location by logging onto the server using a secure login and password. Multiple experts may be awarded simultaneous access and this allows for second opinions and cross-referencing. Since September 2009, the expert has access to the images on his smart phone. The application was created in collaboration with an IT based company — i2i telesolutions, Bangalore and was originally built for the Apple iPhone. The application allows images to be viewed in the similar manner as on the PC and even allows comparison of previous visits. The expert is able to record his findings and create a report using a customized template which is submitted to the server using the GPRS cellular network. (Figure 12.9) This reduces the dependence on the variable speeds of internet and allows the expert to view and report 'on the go'. This technology was voted as one of the Top 10 Medical innovations in 2009 by India Today [13] and featured in the Ministry of External Affairs, Govt of India publication, India Perspectives Jan–Feb 2010 [14].

Fig. 12.8 Project manager uploads the images on to the server.
In the foreground, the project manager uploads the images obtained on a specially hosted server using a customized software for the remote expert to access. In the background, the technicians continue to image other infants.

To improve follow-up to allow serial examination of these infants, several innovative methods were used including the use of cell-phone reminders, personalized report with key family members, progress cards (a 'yellow ROP card'), sharing the daily systemic report of the baby with the mother (over telephone) and meticulous data collection. The number of cell phone users even in rural areas was extensive in areas where the project has been implemented thus far, and has helped in contacting the parents and improving follow-up.

Results

For the purpose of this report, only results relevant to ICT and Tele-ophthalmology have been presented. *(Disease-specific and treatment-related data have been omitted in view of the large non-medical readership.)*

In close to 28 months of the study, 3120 infants were screened in over 6 rural districts of Southern Karnataka in over 23 centres. With a median of 4

Fig. 12.9 The iPhone Tele-ROP application screen shots.
The workflow on the iPhone application developed by i2i Tele-solutions along with Narayana Nethralaya allows the remote expert to view, compare and report real –time using the cellular network without dependence on the internet.

follow-up visits for each infant and over 14 images per session per infant, the KIDROP database currently includes over 1,80,000 images.

The correlation of the algorithm-based-triage-decisions taken ($n = 967$, pilot cohort) by the technicians with those obtained using the gold standard of binocular indirect ophthalmolscopy (BIO) by the ophthalmologist had 100 per cent agreement in all cases that required treatment, in more serious cases of aggressive posterior ROP and in zone 1 disease (closest to the centre of vision, the macula).

With the disease stages moving to the more peripheral zones, the accuracy of the technicians reduced. In Zone 2 posterior there was 90 per cent correlation, in Zone 2 anterior there was 84 per cent correlation and in the most peripheral Zone 3, there was a 69 per cent correlation. The lower accuracy in the peripheral zones was perhaps due to the inability of the wide-field lens to accurately image the ora in all cases. The occurrence of 'edge artifacts' was a leading deterrent. Hence the milder stages of diseases in the peripheral zones were picked up less accurately. However, with increased experience, the technicians were able to image the ora serrata in a greater proportion of cases (Figure 12.10) and their accuracy in these zones too improved [15].

The other benefits of ICT–tele-ophthalmology in this study: Real-time digital imaging of sick infants in several centres and remote expert analysis also helped in picking up several other pediatric retinal conditions such as metastatic endophthalmitis (seven eyes), white-centered haemorrhages leading to diagnosis of systemic conditions such as heart disease and anaemia (23 cases), retinoblastoma (an ocular cancer, in five eyes), 'cherry-red spot' (six eyes), birth trauma (six eyes) and uveitis (nine eyes). This does not include other retinal vascular diseases such as familial exudative vitreo-retinopathy (FEVR), persistent foetal vasculature syndrome (PFVS) and other ocular conditions which were also diagnosed in appropriate cases.

The Retcam images were also used to determine the effect of laser treatment for documenting outcome measures and also for training purposes. 'Skip areas' — inadvertently not treated by the training ophthalmologist — were imaged and used as a teaching tool in the training program.

Expansion of KIDROP: The KIDROP entered into a public-private-partnership between Narayana Nethralaya and the National Rural Health Mission, Min. of Health and Family Welfare, Govt of Karnataka in September

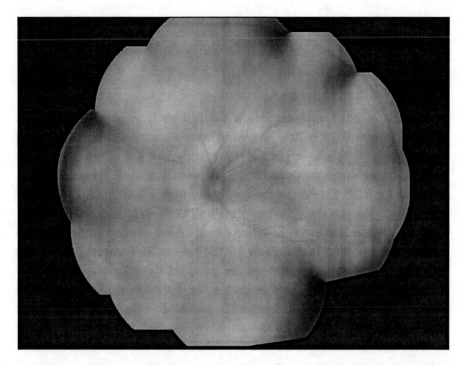

Fig. 12.10 Montage of an infant retina showing 360 degrees of the retina including the periphery. With increasing experience, the ability of the technicians to capture 360 degrees of the peripheral retina has improved along with their accuracy of detecting disease in these more peripheral zones.

2009. The project is poised to expansion into 12 more districts of the state and will include 36–40 additional neonatal care centres. All treatment will be done free of cost under this initiative which will include government and private centres. As part of the PPP, KIDROP has completed free training to Ophthalmologists and Ophthalmic technicians in 9 districts (of the 12) thus far (December 2010). The project is expected to be operational and receive images from the 12 new districts starting from January 2011.

Return on Investment: The KIDROP project (in the past 28 months alone) has screened 3120 premature infants to date and treated (laser) 371 infants. Blindness in these rural infants was prevented at a mean age of 3 months. Each child will live to an average of 65 years (life expectance in India, 2007 data) and earn about Rs 40,000 per annum (Per Capital income, 2009 data). This will save the society and the government 96 crore rupees! This does not include the immeasurable social impact of having 371 *less* blind infants in our community. With the expansion of the KIDROP to 50 plus centres the impact will only increase in the coming years.

Conclusions

To the best of our knowledge this is the first study to investigate the role of tele-ophthalmology in ROP screening in rural and semi-urban areas using non-ophthalmologists to screen and report as the first line of care. Previously published reports from the United States [16] and Germany [17] have used tele-medicine for ROP. They differ fundamentally from KIDROP in: 1) KIDROP uses a single Retcam shuttle to cover multiple centres, 2) Technicians image, store, and *report* live in the peripheral rural centre whereas in the previous studies, the reporting is done only by the remote ROP expert [18]. 3) the reporting is real-time and mothers are provided the diagnosis before they leave the peripheral centre. In other studies, the reports are read retrospectively within a day (or a week) and feedback is provided to the centre for in-house babies alone [19].

This study is ongoing at the time of publication of this report. In our model, unlike elsewhere, we investigated the role of non-ophthalmologists, i.e. trained technicians in screening for ROP. Medico-legal aspects notwithstanding, this model was adopted because it is believed that in a country like ours where resources are limited, there are insufficient experts to provide screening in remote areas. In these places where *no* screening has ever been done, this model that allows a 'trained technician' to screen, may be the answer to the hitherto inadequate coverage of the ROP screening program. The model has been discussed now by the Harvard Business Review as an example of 'reverse innovation' for other developed countries to emulate [20].

There is a social aspect to the study that provides the underlying emphasis for expansion. This includes the fact that over 68 per cent of the infants undergo free treatment and an additional 15 per cent undergo subsidy. The cost-effective analysis of this project is currently under study by Indian Institute of Management, Ahmedabad at the time this report is submitted.

The pilot validation results seem to suggest that trained technicians are capable of accurately (100 per cent) diagnosing treatment-requiring stage of disease, severe disease (including aggressive posterior ROP), and Zone 1 disease. Their ability to detect any stage in the peripheral zones was limited

(69 to 84 per cent). This may suggest that their ability to 'discharge' a baby from follow-up may be limited, since this entails the confirmation of complete vascularization upto the ora serrata. However, as indicated, the ability of capturing the periphery improved with increasing experience. Long-term studies using other technicians are on-going to confirm this finding.

The educational qualifications of the technicians did not seem to play a significant role in their relative accuracies. In the original study, technician 1 was a trained ophthalmic photographer and Technician 2 had not completed high school. Both received the same RETCAM training and hands-on experience. Both showed comparable accuracies. This suggests that like several Ophthalmic disorders which rely heavily on images for diagnoses, ROP too works on the principles of 'pattern recognition'. Experience is thus more valued than previous training. This is the single largest contribution of digital imaging in the field of pediatric retinal management. However, the most important factors for the success of this tele-ROP program are patience, perseverance and passion. Using the principles IEC (i.e. the Triple 'T' approach) and ICT (the KIDROP model), this study illustrates that the problem of childhood blindness in India's rural and under-served areas may be addressed innovatively.

Summary

This report deals with the experience of using a portable wide-field digital imaging device capable of capturing retinal images of neonates as a screening tool for a common and potentially blinding cause of infant blindness (Retinopathy of Prematurity, ROP) prevalent in premature babies. The setting is semi-urban and rural areas of Karnataka state. To the best of our knowledge this is the first study investigating the role of tele-ophthalmology in infant blindness prevention in under-served areas. The report summarizes 28 months experience of the *KIDROP* project (Karnataka Internet Assisted Diagnosis of Retinopathy of Prematurity), pioneered by Narayana Nethralaya Postgraduate Institute of Ophthalmology, Bangalore, India since 2008.

The images are captured, processed, stored and analysed by 'trained technicians'. The accuracy of these technicians was compared with the gold standard of clinical examination by disease experts and the scores evaluated. Images are uploaded using a customized template onto a dedicated server which is accessed by the remote expert on his or her PC or smart phone and the reports are sent back real-time. The paper also discusses the use of ICT in tele-ophthalmology. The long term implications of the study is the possibility of a 'trained technician on site with a remotely situated expert' model that may be used to tackle the scourge of ROP blindness in the under-served areas of our country and perhaps other countries with similar demographics.

The ICT technology used in this project has been voted in the Top 10 medical innovations of 2009 and has been featured by the Ministry of External Affairs (India Perspectives, Jan 2010) as a model for health care delivery to rural areas. Recently (November 2010), the Harvard Business Review featured the project as an example of "reverse innovation".

The content of the report has been modified for a non-medical audience. Adaptations of the report have been previously presented at the World Ophthalmology Congress 2008, the American Academy of Ophthalmology 2008, the Asia Arvo 2009 and the Asia Pacific Association of Ophthalmology 2009 and the American Academy of Ophthalmology 2010 and at the Asia ARVO 2011.

The authors have no financial interest in the product(s) discussed in the report.

For further details please contact the authors.

Please visit the website www.narayananethralaya.org for more details on our community paediatric projects. The video of KIDROP trial (initial 2008) may be viewed on YouTube on the link: http://in.youtube.com/watch?v=brPUh9Zr9zc. The video is titled, 'Towards a World Without Blind Infants'.

Acknowledgements

The authors thank (alphabetically) Sivakumar Munusamy (Ophthalmic Photographer), Krishnan N. (Ophthalmic Photographer) and Praveen Sharma (Project Manager), who have been the centre-stone of the project since its inception and without whom this study would not have been possible.

The participating hospitals in the KIDROP study are (alphabetically): Adichunchinagiri Institute of Medical Sciences, Belur; Bowring Hospital, Bangalore, Cheluvamba Hospital, Mysore, Columbia Asia Referral Hospital, Bangalore; Command Hospital, Bangalore; Chord Road Hospital, Bangalore, District Hospital, Tumkur, ESI Hospital, Bangalore, Gurushree

Hospital, Bangalore, JSS Medical College and Hospital, Mysore; Mandya Institute of Medical Sciences, Mandya; M.S. Ramaiah Memorial Hospital & Teaching Hospital, Bangalore; Narayana Nethralaya Postgraduate Institute of Ophthalmology 1 and 2, Bangalore; Narayana Hrudhalaya, Bangalore; Panacea Hospital, Bangalore; Philomenas Hospital, Bangalore; Siddaramanna Hospital, Tumkur; Siddhartha Medical College, Tumkur; Sri Devaraj Urs Medical College and Hospital, Kolar; Suguna Hospital, Bangalore; Vatsalya Hospital, Mandya.

References

[1] R. Charan, MR Dogra, A Gupta, A. Narang, 'The incidence of retinopathy of prematurity in a neonatal care unit', *Indian J Ophthalmol*, 1995, September, Vol. 43, no. 3, pp. 123–6.

[2] C. Gilbert, 'Retinopathy of prematurity: a global perspective of the epidemics, population of babies at risk and implications for control', *Early Hum Dev*, 2008, Vol. 84, no. 2, pp. 77–82.

[3] *National Neonatalogy Forum of India: National Neonatal Perinatal Database, Report for year 2003–2003*, New Delhi, 2005.

[4] R. Charan, MR Dogra, A Gupta, A. Narang, 'The incidence of retinopathy of prematurity in a neonatal care unit'; A. Vinekar, M.R. Dogra, T. Sangtam, A. Narang, A. Gupta, 'Retinopathy of prematurity in Asian Indian babies weighing greater than 1250 grams at birth: ten year data from a tertiary care center in a developing country', *Indian J Ophthalmol*, September–October 2007, Vol. 55, no. 5, pp. 331–6; R.K. Pejaver, A. Vinekar, A. Bilagi, *National Neonatology Foundation's Evidence Based Clinical Practice Guidelines 2010. Retinopathy of Prematurity* (NNF India, Guidelines) 2010, pp. 253–62.

[5] A. Vinekar, *The ROP challenge in Rural India: Preliminary report of a telemedicine screening model. (In) International Experience with photographic imaging for pediatric and adult eye disease: Retina Physician* 2008; Supplementary 9–10; J. Kreatsoulas, 'Progress in ROP Management through Tele-Ophthalmology', *Retina Today*, November–December 2010.

[6] C. Gilbert, 'Retinopathy of prematurity: a global perspective of the epidemics, population of babies at risk and implications for control', *Early Hum Development*, 2008, Vol. 84, no. 2, pp. 77–82.

[7] 'Early treatment of Retinopathy of Prematurity cooperative group. Revised indications for the treatment of retinopathy of prematurity: Results of the early treatment of Retinopathy of Prematurity Randomized trial', *Arch Ophthalmol*, 2003, Vol. 121, pp. 1684–96.

[8] G. Sanghi, M.R. Dogra, P. Das, A. Vinekar, A. Gupta, S. Dutta, 'Aggressive posterior retinopathy of prematurity in Asian Indian babies: spectrum of disease and outcome after laser treatment', *Retina*, 2009, October, Vol. 29, no. 9, pp. 1335–9.

[9] A. Vinekar, M.R. Dogra, et al., 'Retinopathy of prematurity in Asian Indian babies weighing greater than 1250 grams at birth'.

[10] A. Vinekar, M.R. Dogra, et al., 'Retinopathy of prematurity in Asian Indian babies weighing greater than 1250 grams at birth'; and J. Kreatsoulas Progress in ROP Management through Tele-Ophthalmology.

[11] Photographic Screening for Retinopathy of Prematurity (Photo-ROP) Cooperative Group, 'The Photographic Screening for Retinopathy of Prematurity Study: Primary Outcomes', *Retina*, March 2008, Vol. 28, no. 3 Supplement), pp. 47–54; A. Vinekar, M.T. Trese, A. Capone Jr; Photographic Screening for Retinopathy of Prematurity (PHOTO-ROP) Cooperative Group, Evolution of retinal detachment in posterior retinopathy of prematurity: impact on treatment approach, Am J Ophthalmol, 2008 March, Vol. 145, no. 3, pp. 548–55.

[12] Early treatment of Retinopathy of Prematurity cooperative group. Revised indications for the treatment of retinopathy of prematurity: Results of the early treatment of Retinopathy of Prematurity Randomized trial. Arch Ophthalmol, 2003, Vol. 121, pp. 1684–96; M.R. Dogra, A. Vinekar, K. Viswanathan, T. Sangtam, P. Das, A. Gupta, S. Dutta, 'Laser treatment for retinopathy of prematurity through the incubator wall: Ophthalmic Surg Lasers Imaging', 2008 July–August, Vol. 39, no. 4, pp. 350–2.

[13] Top Ten Medical Innovations: iPhone used to stave off blindness. India Today, 28 December 2009, pp. 126–30.

[14] *A Unique Experiment in Tele-Medicine: Tele-Ophthalmology provides a new hope in preventing infant blindness in rural India Perspectives*, Ministry of External Affairs, Govt. of India, Vol. 24, no. 1, 2001, pp. 70–1.

[15] A. Vinekar, The ROP challenge in Rural India: Preliminary report of a telemedicine screening model. (In) International Experience with photographic imaging for pediatric and adult eye disease; J. Kreatsoulas, Progress in ROP Management through Tele-Ophthalmology.

[16] R.A. Silva, Y. Murakami, A. Jain, J. Gandhi, E.M. Lad and D.M. Moshfeghi, 'Stanford University Network for Diagnosis of Retinopathy of Prematurity (SUNDROP): 18-Month Experience with Telemedicine Screening', *Graefes Arch Clin Exp Ophthalmol*, January 2009, Vol. 247, no. 1, pp. 129–36.

[17] B. Lorenz, K. Spasovska, H. Elflein, N. Schneider, 'Wide-field digital imaging based telemedicine for screening for acute retinopathy of prematurity (ROP): Six-year results of a multicentre field study', *Graefes Arch Clin Exp Ophthalmol*, September 2009, Vol. 247, no. 9, pp. 1251–62.

[18] Silva, R.A., Y. Murakami, A. Jain, J. Gandhi, E.M. Lad and D.M. Moshfeghi, 'Stanford University Network for Diagnosis of Retinopathy of Prematurity (SUN-DROP): 18-Month Experience with Telemedicine Screening'; B. Lorenz, K. Spasovska, H. Elflein, N. Schneider, 'Wide-field digital imaging based telemedicine for screening for acute retinopathy of prematurity (ROP): Six-year results of a multicentre field study'.

[19] Ibid.

[20] A telemedicine innovation for the poor that should open eye. Harvard Business Review 7 November 2010. http://blogs.hbr.org/govindarajan/2010/11/a-telemedicine-innovation-for-the-poor-that-should-open-eyes.html.

13

Developing Hardware-Software Suite for AMAL-Adaptive Modular Active Leg

G.C. Nandi, Auke Jan Ijspeert*, Varun Soneker
and Anirban Nandi

With the advancement in storage and retrieval capacity of computers and the availability of inexpensive but powerful microcontrollers, it is now possible to develop most challenging above-knee active prosthesis with environmental adaptation providing maximum physical and psychological comfort to an amputee. In this paper, we propose the use of robotics and AI (Artificial Intelligence) techniques, specifically those of biped locomotion for use in active prosthesis. First an overview of a new Adaptive Modular Active Leg (AMAL) has been presented followed by the detailed control strategy for generating active damping profile implemented through a microcontroller for execution-level control of a specially configured Magneto-Rheological (MR) damper. Unlike several active prosthetic legs currently available, which are designed as stand-alone devices, based on sensory feedback and rigid control, the present design is heavily integrated with biological motor control circuits available with the amputee. The technology developed so far has been retrofitted on a professionally made prosthetic device called AMAL. The system is characterized by its simplicity, robustness and reliability.

Disability Rehabilitation Management through ICT, 111–138.

Introduction

The literal meaning of the word 'amputee' is 'unfortunate person'. Does fortune depend on the physical ability of a person? It is commonly observed that disabled persons have limited access to education, employment and basic health care. The civilized society has enacted many laws against discrimination of the disabled, created special arrangements in official building constructions, in airports and other places and some facilities in mass transport systems. Still, by and large, disability is viewed as a burden in the social environment around the sufferer, in spite of the fact that amputations are mainly caused by accidents, diseases and congenital anomalies over which an individual has no control or responsibility. Approximately 75 per cent amputations are due to peripheral vascular diseases caused by poor circulation of blood, and cancer; 22 per cent are due to accidents; and 3 per cent are due to birth defects. As noted by Alvin Muildenburg and Bennett Wilson Jr. [1], out of the accident-related amputations, 80 % are suffered by soldiers in land-mine explosions and other wartime duties. Then why do we have such psychological discrimination? Can this discrimination be removed by enacting laws alone? The answer is no. The real discrimination can be removed only if technology can offer them as close a simulation of normal life as possible. The current research is dedicated to millions of such poor and unfortunate fellows of ours with an objective of giving them a normal walking life in all fronts — slow walking, brisk walking, stair climbing, and running, to name a few. There are more than two amputees per thousand in the Indian subcontinent, i.e. approximately 2.5 million people. About 1.5 amputees per thousand live in the USA and Canada [2].

Above-knee (trans-femoral) amputees form the second largest group of amputees in the world. However, the challenges for solving the above-knee prosthesis problem are manifold. But the advancement of computational speed, storage and retrieval facilities coupled with the availability of better tools for designing necessary software-hardware integration, advancement of biomedical engineering, cognitive robotics and Central Pattern Generators (CPG) are showing some hope.

Many significant research works in above knee prosthesis, especially passive prosthesis, had been initiated by J. Wallach [3], J.B. Morrison [4], W.C. Flowers of MIT [5], G.W. Horn [6], L. Disano et al. [7], and Max Donath [8]. Computer-controlled above-knee prosthesis research began at the

Massachusetts Institute of Technology (MIT) in 1969 mainly with the initiative of Prof. Woodie Flowers. Prof. Flowers designed an electro-hydraulic simulator. However, its main limitation was that the amputee needed a handheld computer for real-time damping control. Almost a decade later, in 1978, Mark Tanquary devised a passive prosthesis using MC6800 processor. These classical works translated into various prosthetic devices (mainly passive) which are currently available.

However, all the contributions that laid the foundation of research in prosthesis were made in the early or late 1970s when information technology and computer-related hardware were crude and expensive as compared to the present day. With the increase in speed, storage and retrieval capacity of computers, many researchers started working on algorithm-based approaches, with either trajectory- or torque-based control. D. Zlatnik [9] used the soft control non-analytical technique. Kalanovic et al., [10] worked on the feedback error learning neural network scheme. Grimes [11] worked with echo control scheme based on input from a biological leg. More recently, Ari J. Wilkenfeld (PhD thesis, MIT, 2000) worked on biologically-inspired control of knee prosthesis. At present a number of semi-active and semi-intelligent legs have been developed and marketed by companies like Otto bock, Endolite and Victhom [12]. However, in all these devices the technology development has taken place taking prosthesis as a stand-alone device and those are expensive, some costing tens of thousands of US dollars, making them unsuitable for use by common people, especially in the developing countries of South-east Asia and Africa. Moreover, these companies are careful not to reveal their technology and thereby maintain a monopoly over such products.

Secondly, the availability of cost-effective microcontrollers and adaptive actuators (like magneto-rheological damper or MR damper) encourage us to take up the challenge of creating suitable intelligent prosthesis, which can give maximum comfort to the patient. Such systems can give almost normal walking life and remove the embarrassment of losing a leg or using a passive leg. Keeping the common man's suffering in mind, the authors have undertaken extensive research work towards the development of a software-hardware suite and its integration with an intelligently adaptive, active prosthetic leg.

The technologies used are derived from the experiences gained from extensive research in the area of robotics, artificial intelligence and control. To make the system inexpensive, sensor-based feedback has been kept to a minimum.

This paper first gives an overview of the development of AMAL; section two discusses a low-cost active prosthesis prototype; section three discusses the methodology of capturing and providing most appropriate stable gait pattern for an amputee; and the last section discusses the damping control strategy with necessary hardware implementation for AMAL.

Overview of Amal (Adaptive Modular Active Leg)

A. *Human Gait Cycle*: Human walking is a coordinated effort of neuromuscular and musculoskeletal structures and the capsular constraints of the joints, as well as the energy generated by the upper part of the human body while swinging the hands. Every human being has a unique walking pattern which too depends on the mode he is in, that is, whether he is walking, running, jumping, and so on. So, to reduce the complexity of this cycle the gait has been divided into various phases.

The gait cycle is the period of time between any two identical events in the walking cycle. A complete gait cycle can be viewed in terms of three functional tasks of weight acceptance, single-limb support and limb advancement. The gait cycle can be described in the phasic terms of initial contact, loading response, mid-stance, terminal stance, pre-swing, initial swing, mid-swing and terminal swing (Figure 13.1).

(i) *Initial contact* is an instantaneous point in time only and occurs the instant the foot of the leading lower limb touches the ground.

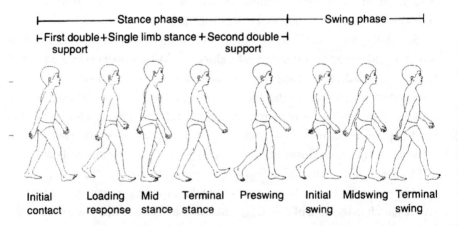

Fig. 13.1 Human Gait Cycle Phases (http://www.nupoc.northwestern.edu/prosHistory.shtml).

(ii) *Loading response*, the foot comes in full contact with the floor, and body weight is fully transferred onto the stance limb.

(iii) *Mid-stance* begins when the contralateral foot leaves the ground and continues as the body weight travels along the length of the foot until it is aligned over the forefoot.

(iv) *Terminal stance* constitutes the second half of a single-limb support. It begins with heel rise and ends when the contralateral foot contacts the ground.

Pre-swing begins when the contralateral foot contacts the ground and ends with the ipsilateral toe off. During this period, the stance limb is unloaded and body weight is transferred onto the contralateral limb.

Initial swing begins the moment the foot leaves the ground and continues until maximum knee flexion occurs, when the swinging extremity is directly under the body and directly opposite the stance limb.

Mid-swing is a critical event which includes continued limb advancement and foot clearance. This phase begins following maximum knee flexion and ends when the tibia is in a vertical position.

Terminal swing is the phase in which the tibia passes beyond the perpendicular and the knee fully extends in preparation for heel contact. For more details about this one can see [24].

B. *Amputee Gait Cycle:* So far all artificial leg designers have tried to develop the limb as a stand-alone device equipped with multi sensor-based control strategy for producing matching gait patterns that would allow the amputee to walk in the most efficient manner. However, they have completely ignored the sophisticated biological motor control circuits available for integration at the subsequent higher joints, for example, hip joints, shoulder joints, etc., for an above-knee amputee.

Our focus is on tapping maximum from the biological CPG (Central Pattern Generator) of the amputee and to try to solve a number of challenges arising out of it.

Hardware Description of AMAL

This section will provide a brief overview of AMAL on which our technology has been retrofitted.

Only those aspects necessary for explanation of the control will be elaborated.

An above-knee amputee uses a prosthesis system which contains the following basic units:

- The socket
- The knee fitted with sensor
- Foot-ankle unit

In AMAL all parts are modular and made very professsionally by ALIMCO (Artificial Limb Manufacturing Corporation, India). AMAL hardware details are shown in Figure 13.2.

In a successful prosthesis, the fitting of the socket to the stump is a very important factor. In the amputee's first visit to ALIMCO, an impression of the sockets is taken from the remaining part of the amputee's leg stump and the prosthetic is designed appropriately (using a CAD software package).

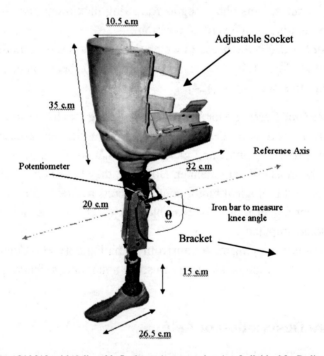

Fig. 13.2 Parts of AMAL with Adjustable Socket to Accommodate Any Individual for Preliminary Testing.

The knee unit bolts to the bottom of the socket. Apart from socket unit fit, the length of the remaining leg stump is very crucial for the comfort of an amputee. A longer stump generally means a better lever arm to control. Here the role of the surgeon who performed the surgery is very significant. For AMAL the ankle-foot units have been designed effectively — there is a softer ankle built into the rubber foot to enable it to absorb some shock at heel strike.

In AMAL, the knee is actuated by a magneto-rheological damper (MR damper), as shown in Figure 13.3. It consists of magneto-rheological fluids consisting of stable suspensions of micro-sized, magnetizable particles dispersed in a carrier medium. When an external magnetic field is applied, the polarization induced in the suspended particles results in magneto-rheological effect of the MR fluids. This effect influences the mechanical properties of the fluids. The suspended particles in the MR fluids become magnetized and align themselves like chains in the direction of the magnetic field. The formation of these particle chains restricts the movement of MR fluids, causing the increase in viscosity of the fluids and increase in damping. Since the change of damping could be caused by direct influence of the magnetic field, at present it seems to be the most suitable actuator from the point of view of ease of control. Unlike robotic applications electrical motors and other actuators are ruled out due to their power hungriness.

In the AMAL prosthesis, the MR damper, which is a hydraulic damper, provides comfort to the amputee by providing flexion and extension resistance during swing and thus helps keep the swinging of the leg under better control of the amputee.

The MR damper used for AMAL has the following specifications:

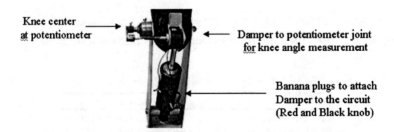

Fig. 13.3 Knee Sensors and MR Fluid-based Actuator for AMAL.

Electrical

Input current-1 Amp max (continuous)

Input volage-12 V DC

Resistance – 5 Ω at ambient temp

Mechanical

Damping force (Peak to Peak) –>= 2224 N

Maximum tensile load >= 4446 N

Maximum operating temperature −71° C

The MR damper has less than 25 ms response time. It can supply continuously variable, speed-independent torque and is fairly insensitive to temperature, which is good for prosthesis damping control. It is smooth, can accommodate realistic force feedback with high fidelity feel, has high torque density (at least 10 times greater than DC motors), is compact and has inherent stability. Some physical dimensions of the leg have been given in Table 13.1.

The schematic diagram of the model, with actuators is presented in Figure 13.4.

Goals of AMAL Prosthesis

- To implement adaptive control strategy with minimum intervention from sensory feedback.
- To provide stability in the early stance phase so that the knee does not get locked.
- To ensure that it proceeds smoothly into the flexion (i.e. swing-flexion) from pre-swing.
- To allow optimum rise of the heel during swing-flexion.
- To facilitate a soft and comfortable stop.

The first goal has been achieved by developing a suitable software framework integrating with a microcontroller and has been described in section 4.

Table 13.1. Dimensions of Prosthesis.

Property	Dimension
Peak knee flexion angle	10°s with respect to horizontal axis
Peak knee extension angle	90°s with respect to horizontal axis
Height of prosthesis	65 cm
Weight of the prosthesis	4.5 kg

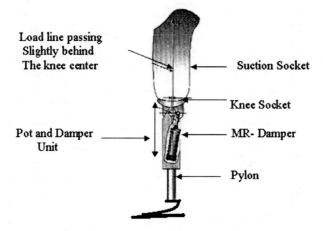

Load line passing
Slightly behind
The knee center

Suction Socket

Knee Socket

Pot and Damper
Unit

MR- Damper

Pylon

Fig. 13.4 The Actuator Fitted into the Feet and Knee Socket.

The second and third goals have been achieved by a suitable mechanical design which allows the altering of the static alignment of the amputee's weight line with respect to the rotation axis of the knee. The fourth one has been accomplished by integrating a special actuator-magneto-rheological damper equipped with special feet (refer to Figure 13.4).

Correspondence Between Control Strategy and State of Walking Gait

After selecting a suitable gait pattern we have developed a control strategy to extract the required damping profile from it. Logically we can see that the amount of damping required for each state of the walking gait cycle changes in accordance with the amount of load that a leg has to bear in that particular state. During stance flexion of a particular leg, it has to bear the maximum load since the weight of the entire body has to be borne by it while the other leg is free to swing. So we see that the damping force in Figure 13.5 is greatest in and around the state of stance flexion. Then as the weight of the body gradually gets shifted to the other leg we find that the damping force goes on decreasing till around stance extension when the entire weight has been shifted to the other leg. Then as the leg starts to swing, i.e. around swing flexion, the damping force is nearly zero because the leg does not have to bear any body weight. Then, at the end of the gait cycle, i.e. during swing extension, the

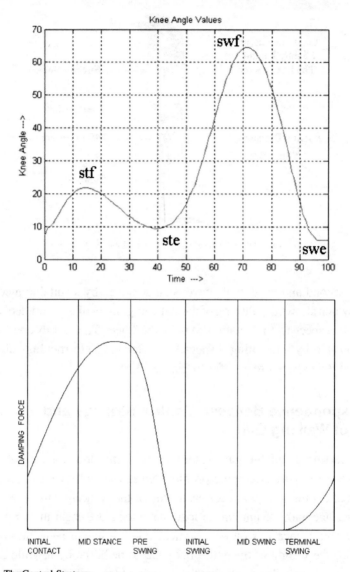

Fig. 13.5 The Control Strategy.
Note: stf stands for stance flexion, ste for stance extension, swf for swing flexion, and swe for swing extension.

damping force again increases in order to make sure that the leg straightens out completely just before heel strike of the next gait cycle. This is how the damping varies throughout a complete gait cycle.

Damping Control Strategy

The control strategy of AMAL is conceptually different from all existing prosthetic legs in the sense that it integrates biological motor control circuits with the prosthetic leg control. It enables us to control the system with less sensory information utilizing biological entrainment from the other joints. This controller is particularly active in the steady state and remains active for long walking periods. Our controller possesses the ability to deal with any abrupt change in gait by immediately making the damping very high so that the leg becomes locked and temporarily passive to ensure maximum safety to the amputee.

Strategy for Damping Profile Generation

Here we have expressed the force to be applied by the damper as a function of the state of the knee (i.e. the knee angle). As shown in Figure 13.6, for stable walking, the force applied by the damper must be maximum when the prosthetic leg is in mid- to late-stance phase (and the biological leg is in the swing phase). This is because at this state the entire weight of the body is borne by the prosthetic leg and the biological leg is off the ground. Then the

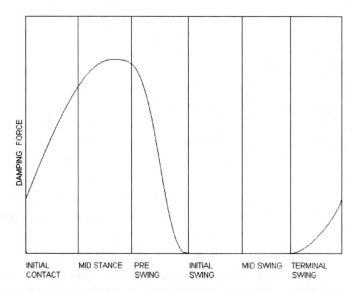

Fig. 13.6 Graph Showing the Damping Force Variation during Various States of Gait Cycle.

force applied by the damper goes on decreasing as the hip senses heel contact of the biological leg and the prosthetic leg approaches initial swing, because the weight of the body is getting gradually shifted to the biological leg and the damper need not provide much force. This decrease goes on until the prosthetic leg leaves the ground. While it swings, the initial and mid-swing require no force from the damper as the main job in these phases is done by the hip muscles, but as the prosthetic leg approaches the end of mid-swing, it becomes necessary to apply some force to straighten out the knee again just before heel strike, otherwise serious injury/discomfort may be experienced by the amputee due to loss of balance.

Hence, we see (in Figure 13.10) a rise in the damping force during terminal swing and this rise in the force continues into the next cycle of the gait. That is why we see the damping force in the initial contact phase starting not from zero but from the point where it had been left at the end of the terminal swing of the previous cycle. The control action has been illustrated in Figure 13.7

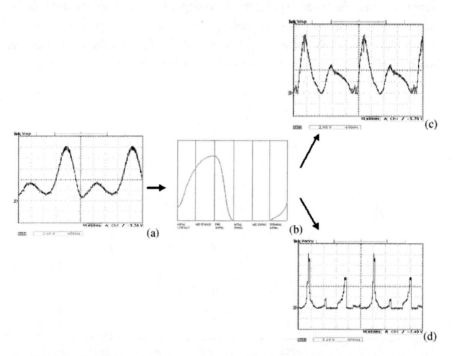

Fig. 13.7 Input-Output Data: A — Knee trajectory as input to our controller; B — Control strategy developed for damping; C — Knee moment obtained corresponding to A; D — Damping profile created corresponding to A.

with knee profile as an input and knee moment and damping profile as outputs.

Implementations

Laboratory testing and implementations have been carried out [13]. The present controller takes up to three steps for initiation of control actions (these three steps the amputee needs to walk as if it is a passive leg). First we decide what walking pattern (simple walking, brisk walking, etc.) to be executed. Then the corresponding damping profile is generated by the controller which is already stored in the form of a look up table. The input-output data are shown in the Figure 13.8. In the subsequent sections only the framework of implementation will be presented.

The whole system is divided into three main parts the Sensory Unit, the Processing Unit and the Actuation Unit, as shown in Figure 13.9.

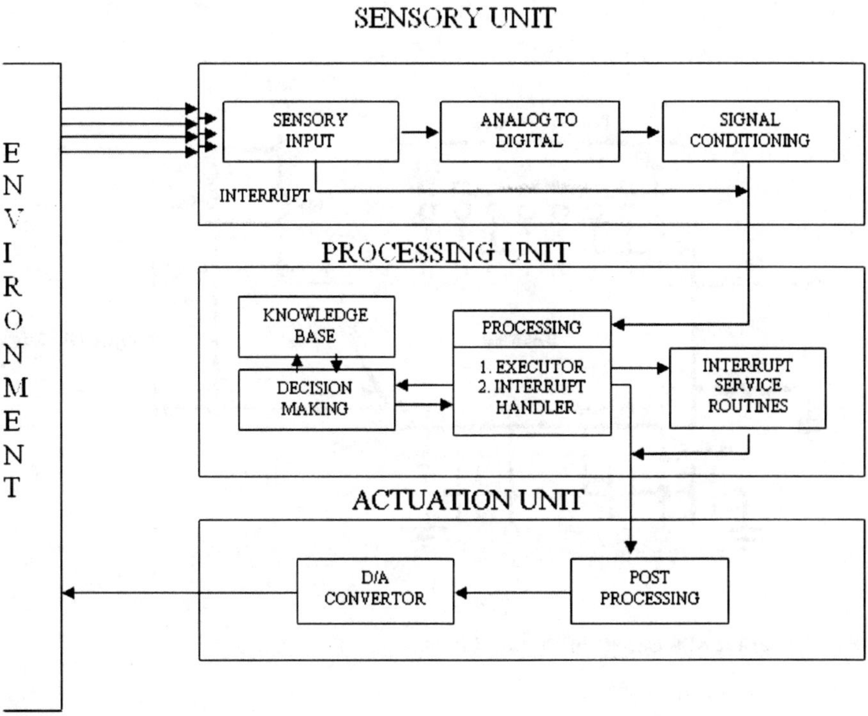

SENSORY UNIT

PROCESSING UNIT

ACTUATION UNIT

Fig. 13.8 System Overview.

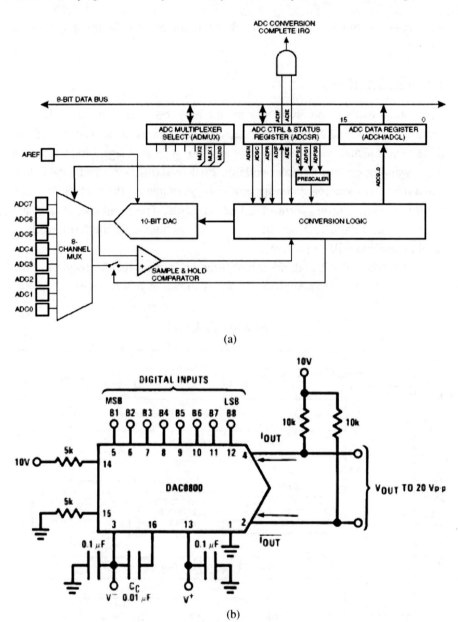

Fig. 13.9 (a) The ADC Circuit. (b) The DAC Circuit.

The Sensory Unit

The sensory unit deals with the sensory values; it collects information from the hip potentiometer and directs them for digital conversion. These sensory data help in knee profile identification from hip profile and accordingly the damping profile data is read for implementation (refer to Figure 13.9). The digital-to-analogue sub-module is just a D/A converter, from where the digital output is passed on for signal conditioning. This is the place where various noise factors from the signal are eliminated with the help of the Kalman Filter. In case the input to this module is an interrupt generated by any sensor, the input is bypassed on to the output towards the processing unit without A/D and signal conditioning because it will be a single bit interrupt.

The Processing Unit

This is dedicated for execution and interrupt handling. Execution may have to take certain decisions while processing the inputs and that is done by decision sub-modules and the result is sent back to the processing sub-module. The decision sub-module consists of a lot of stored information (knee profile vs. damping profile mapping; refer to Figures 13.11 and 13.12) to take decisions, for which there is a knowledge base (whether the current profile corresponds to slow or brisk walking, and so on) directly attached to the decision module (for generating corresponding knee damping profile data) from where the information could be retrieved and new information inserted. In case the input is an interrupt, it is handled by the interrupt handler sub-module, which selects the appropriate Interrupt Service Routine (ISR) and executes it. By suitable interrupt handling we can control any emergency situation. For example, if the person is stopping (which could be sensed within some threshold limit say, 200 ms), and if there is no change in the controller's designated memory locations for capturing profiles, then the damping would be set to maximum, thereby locking the knee and making it passive, so that amputee could be saved from falling down. Finally the output travels to the Actuation Unit.

The Actuation Unit

This unit is oriented towards handling the outputs from previous modules and transforming them into a format that could be sent to the actuator.

Post-processing is to rescale the output so that it could be transformed without any information loss. The D/A conversion is the final module from where the analogue outputs are projected into the actuators in the environment.

Hardware Implementation

Hardware Components Used in the Prosthesis

(1) *Microcontroller*: The AT90S8535 is a low-power CMOS 8- bit microcontroller based on the AVR RISC architecture [14]. The AVR core combines a rich instruction set with 32 general-purpose working registers, which have been very beneficial to hold the values of different parameters at the same time in the prosthesis. The AT90S8535 provides the following features: 8 K bytes of In-System Programmable Flash, 512 bytes EEPROM, 512 bytes SRAM (Flash memory has been used to store the main program and EEPROM is used to simulate the look up table inside the microcontroller). AT90S8535 has eight channels inbuilt, and a ten-bit ADC, through which more than one input can be taken at the same time. And since there are eight channels, there could be at most eight inputs only. The ADC provides the resolution of ten bits which increases the precision of the input signals. The microcontroller provides four ports which could be programmed as input or output by setting the data direction register. Individual pins of the ports can also be programmed as input or output. One of the ports has an additional facility to take eight different analog inputs and it supports the inbuilt ADC.

(2) *Analog-to-digital converter*: The AT90S8535 features a ten- bit successive approximation ADC [15]. The ADC is connected to an eight-channel Analog Multiplexer (as shown in Figure 13.10a).

The ADC converts an analog input voltage to a ten-bit digital value through successive approximation. The analog input channel is selected by writing to the MUX bits in ADMUX. Any of the eight ADC input pins ADC 7.0 can be selected as single-ended inputs to the ADC.

The ADC can operate in two modes–single conversion and free running. In the single conversion mode, each conversion will have to be initiated by the user. In the free running mode, the ADC constantly samples and updates the ADC data register. This enables the system to be adaptive. The ADFR bit in the ADCSR selects between the two available modes. We initially used

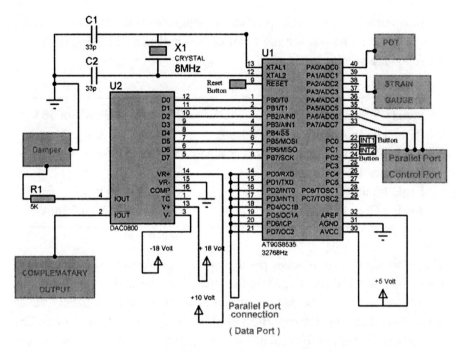

Fig. 13.10 Overall Interfacing Diagram.

the free running mode to ensure that the system got the updated values of the sensors and the system worked in real time, But we found that most of the time was spent in Interrupt Service Routine for the interrupt generated by ADC on free running mode and that reduced the response time of the microcontroller, which is why we moved to the single running mode.

Conversion starts by writing a logical "1" to the ADC Start Conversion bit, ADSC. This bit stays high as long as the conversion is in progress and will be set to zero by the hardware when the conversion is completed. The ADC generates a ten-bit result, which is presented in the ADC data registers ADCH and ADCL. When reading data, the ADCL must be read first, then the ADCH, to ensure that the content of the data register belongs to the same conversion. Once the ADCL is read, the ADC access to the data register is blocked. Then the ADCH is read, and ADC access to the ADCH and ADCL registers is re-enabled.

(3) *Digital-to-analog converter*: The DAC0800 is used to convert eight-bit digital data into analog output, supplemented towards the MR damper [19]. This

is a monolithic eight-bit high-speed current-output digital-to-analog converter (DAC) featuring typical settling times of 100 ns (Figure 13.10b) and having the following specifications:

Supply Voltage	(V+ to V−) ±18 V or 36 V
Reference Input Differential Voltage	(V14 to V15) V− to V+
Reference Input Common-Mode Range	(V14, V15) V− to V+
Reference Input Current	5 mA
Logic Inputs	V− to V+ 36 V
Analog Current Outputs	(VS− = −15 V) 4.25 mA

The reference voltage is set as per the required conditions. The eight bits of microcontroller is directly fed into the DAC at B1–B8 and the analog output I_{out} is used to control the damper. Power is supplied from the battery via an appropriate regulator. The capacitors used in voltage supply are for controlling the voltage fluctuation. If there are no voltage fluctuations the capacitors can be removed from the circuit.

(4) *Sensor used*: We have used Potentiometer as a sensor. It has been used to measure the potential in a circuit by tapping off a portion of a known voltage from a resistive slide wire and comparing it with the unknown voltage by means of a voltmeter. A 5 K wire wound-potentiometer has been use in the hip. Wire-wound potentiometers provide bigger housing diameter resulting in flexibility and good linearity.

(5) The Actuating Device: Here we have used an MR damper manufactured by LORD Company, USA. The magneto-rheological (MR) damper is a semi-active control device that is capable of generating the magnitude of forces necessary for full-scale applications required in the prosthesis. The MR damper produces the required variable resistance generated from the above mentioned software-controlled microcontroller for the knee movement. With the controlled signal, when a magnetic field is applied to the MR fluid inside the mono tube housing, the damping characteristics of the fluid increase with practically infinite precision and in under 25 ms response time.

Third Party Software Used

The coding is done in Assembly language as well as C language; the programs were assembled and compiled with the help of compilers best suited for the program. The following compilers were used in the procedure:

1. *CodeVisionAVR C Compiler V1.24.9:* High Performance C Compiler, Integrated Development Environment for AVR Applications, available for download with limited 2 k bytes of memory [16].
2. *AVR Studio 4.12.460:* GUI Version, Integrated Development Environment (IDE) for writing and debugging AVR applications, available for download [17].
3. *Fast AVR ver.4.1.3:* Basic Compiler for AVR, this compiler was used initially [18].

The overall schematic diagram of interfacing of sensors, actuator and microcontroller (Figure 13.11)

Circuit Details: The circuit diagram represents overall sensor, actuator and microcontroller interfacing. POT is a 5 kilo ohm potentiometer attached at first pin (PA0/ADC0) of PORT C in the microcontroller. Similarly the strain gauge is attached at the second pin (PA1/ADC1) of the same port. Only this port could be used to take analog inputs according to the specification of the microcontroller. And they could be used either for analog input or digital input (0 or 5 volts). The controller needs to be configured for using the pins as analog input pins.

An 8 MHz crystal oscillator is attached across the XTAL1 and XTAL2 pins of the microcontroller to provide clock pulse for the microcontroller with two capacitors of 33 pf each attached to both legs of the oscillator across Ground. Three mechanical buttons are used in the circuit; one is the reset button that brings the whole circuit in the initial state while the remaining two of them are dedicated to user-generated interrupts whenever one wishes to switch between the modes.

All eight pins of PORT D of the microcontroller are connected with the data pins of the parallel port, and PA5, PA6 and PA7 pins of Port A are connected with the control pins in the parallel port. The controller is operated at 5 volt, across a common ground throughout the circuit. Eight pins of PORT

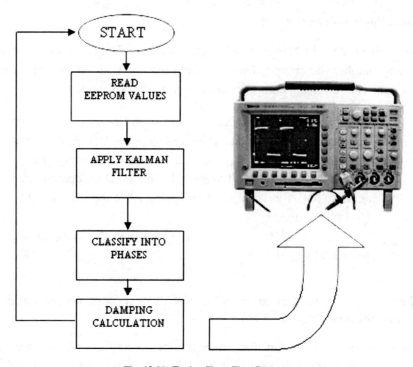

Fig. 13.11 Testing Phase Flow Diagram.

B are directly connected to the eight input pins of the DAC which works at a power supply of -18 to $+18$ volt. Reference Voltage is kept 10 volt, so the output at pin I_{out} varies between 0 and 10 volt, depending on the pattern of input applied. Complementary Output is the residual current that is left at $I_{out} (\bar{I}_{out} = V_{ref} - I_{out})$.

Testing System Performance on Test Data

The testing is very important to check that the hardware and software have been integrated properly.

The knee angle and moment values are taken from existing datasets. Those dataset values are stored into the EEPROM inbuilt in the microcontroller. Later on, the corresponding damping profile (refer Figure 13.9) are fetched directly from the memory and passed on to the processing unit. The processed data gives the damping magnitude values which are passed on to the Digital Storage Oscilloscope. The flow of testing method is shown in Figure 13.12.

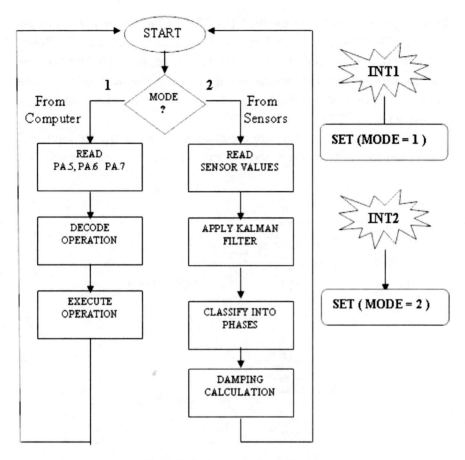

Fig. 13.12 Implementation Flow Diagram.

Writing EEPROM: EEPROM needs to be written with the test data. Initially these data have been supplied from the knee potentiometer and strain gauges. A C program is written to accomplish that. A definite space is reserved in the EEPROM and a pointer is maintained to refer that location and later those reserved locations are filled with the data via those referencing pointers.

Example: eeprom Array[n] = {a, b, c, d........};

where n represents the number of test data samples for joint angles (100 samples from a single gait cycle); a, b, c, are the values of those data. CodeVision-AVR is used to compile the C file. The compiled output produces a separate file for EEPROM with extension .eep. Similar separate files are created for knee trajectories obtained from the CPG. That EEPROM file is programmed into the EEPROM in the controller. Thus the values are stored in EEPROM.

Methodology for implementing the system on the prosthesis. The data from the sensor is incorporated into the system and the performance of the system is verified on the real world data. This phase includes parameter updating as shown in Figure 13.13.

ADC facility is provided with the microcontroller controller itself. A/D conversion runs in two modes: the free running and the interrupt mode. The

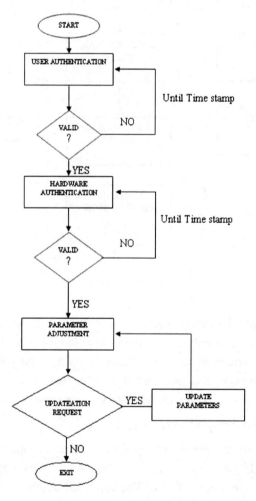

Fig. 13.13 Software Flow Diagram.

first mode starts the new A/D conversion as soon as the converted data is read from the registers holding the results. The second is where an interrupt is generated after a conversion and next conversions do not start until it is asked to do so by setting a bit in its configuration register. This degrades the response time of the main controller program. On the other hand, in the interrupt mode, the service routines will be executed a lesser number of times.

Software Development Methodology

Overview of the System

The software developed using Java 1.4 is made to communicate with the microcontroller via a parallel port. The main task of the software is to provide an interface to the user through which one can enter the control parameters into the microcontroller via the parallel port. The software also provides user authentication and hardware authentication features for security purposes, according to which the user must have a valid user id and a unique password, and along with that the correct hardware must be connected to the software to proceed, or else the software will be blocked. The software provides the facility of downloading the control parameters to the microcontroller. The control parameters are damping profile values shown in Figure 13.8. For persons having different weight and height, the user just needs to set the values with the sliders in the GUI for setting the upper and lower bounds of the corresponding phases and then press the update button upon which a suitable gait pattern would be configured and would transfer all the parameter values into the microcontroller.

The software interfacing is possible only when the microcontroller is working in MODE 1. The software works in interrupt-oriented fashion where the interrupts are generated by the user and the microcontroller responds accordingly by taking appropriate actions. The software flow diagram has been shown in Figure 13.14.

The program starts with the user authentication process which checks the log-in and password of the user. Then the program proceeds towards hardware authentication wherein the program sends a byte to the microcontroller by selecting a definite configuration of control pins (for control pins, refer to the next section). Then the program waits until the microcontroller responds with an ID until the time limit expires for authentication. Once the valid id is detected the user is allowed to enter into the software, where he is provided with a form displaying the various modifiable parameters in the form of sliders. The user sets the upper and lower bounds of the damping profile for amputees having various characteristics (height, weight). The damping profile is scaled up/down in the microcontroller when these settings are sent to the microcontroller. The user can repeat the procedure by adjusting the new parameter values or he can exit from the software by just pressing the 'exit button'.

Testing for hardware-software integration. The methodology described here has been implemented on the hardware (microcontroller) using Assembly and C languages. The results obtained in the testing phase were in the form of waveforms which were captured on the Digital Storage Oscilloscope (DSO) and are shown in Figures 13.14-1–13.14-4.

Initially the graphs for knee angle for test data were plotted on the Matlab, which is shown in Figure 13.14-1. Later on those same values were used to produce waveforms as the output of the microcontroller and that result is shown in Figure 13.14-2.

Since the test data that we have used does not contain noise and in actual implementation there will be noise associated with the knee/hip potentiometers some noise was randomly introduced into the test data to check

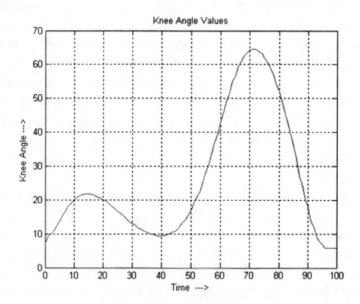

Fig. 13.14-1 Knee Angle Graph Plotted on Matlab.

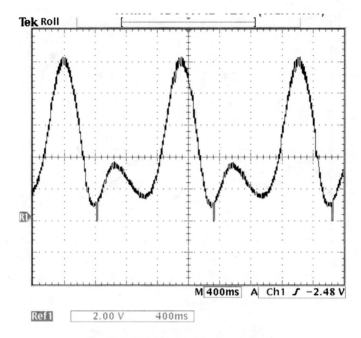

Fig. 13.14-2 DSO1 Actual knee angle Graph.

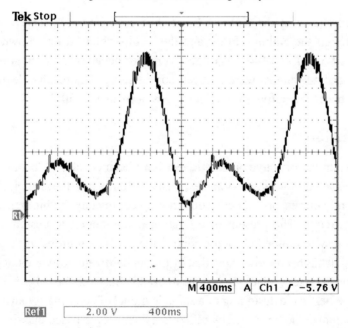

Fig. 13.14-3 DSO3 Eroded knee angle Graph.

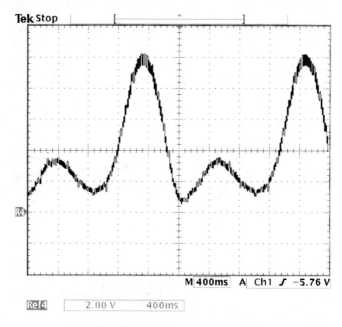

Fig. 13.14-4 DSO5 Corrected knee angle Graph.

the performance of the Kalman Filter. The corrupted values were made to pass through the Kalman Filter and the results obtained are shown in Figures 13.14-3 and 13.14-4. Spikes that were introduced by random noise in the graphs for knee angles were reduced considerably. The results clearly show that the hardware-software integration has been successful.

Conclusion

So far a number of active prosthetic legs have been developed, but all of them behave as stand-alone devices and expensive. Mounted on the artificial leg they heavily rely on sensors and control, ignoring completely the biological motor control circuits. The present investigation is an attempt to design a system integrated with biological motor control that exists within us in the form of coupling. It has been shown here that by proper hardware-software integration, many control issues related to prosthesis could effectively be solved. The main challenge lies in the designing of a suitable model of coupling for an amputee. The problem could be solved by utilizing a Fuzzy logic-based classifier which can provide a reasonably good initial gait pattern for an amputee. The pattern could then be used as a foundation for the reverse engineering of a CPG design for an amputee. Once designed properly, it could be used as a signal generator for control feedback with minimum intervention from the sensors. Since, for active prosthesis, energy plays an important role while a human being is walking, future research could be directed towards analyzing the energy exerted by the coupling actions of other parts of the body and how to tap it suitably and integrate it with the actuator. This would make the prosthesis energy-efficient and less expensive.

Acknowledgement

This research was done under Indo-Swiss collaboration. The authors are thankful to the Swiss National Science Foundation and Department of Science and Technology, Government of India, for their support and funding for carrying out this research.

References

[1] Muildenburg, Alvin L. and A. Bennett Wilson Jr., 'A Manual for Above Knee Amputees', 1996. http://www.oandp.com/resources/patientinfo/manuals/

[2] Ibid.

[3] Wallach, J., 'Control Mechanisms Performance Criteria for an Above Knee Leg Prosthesis', Ph.D. thesis, Rensselaer Polytechnic Institute, 1966.

[4] Morrison, J.B, 'The Mechanisms of Muscle Function in Locomotion', *Journal of Biomechanics*, Vol. 3, 1970, pp. 431–51.

[5] Flowers, Woodie C., 'A Man-Interactive Simulation System for Above Knee Prosthesis Studies', Ph.D. thesis, MIT, August 1972.

[6] Horn, G.W., 'Electro-Control: An EMG-Controlled Above Knee Prosthesis,' *Medical and Biological Engineering*, Vol. 10, 1972, pp. 61–73.

[7] Disano L., et al., 'An EMG Controlled Knee Locking Prosthesis', *Abstracts: Second Annual New England Bioengineering Conference*, 29–30 March 1974, Worcester Polytechnic Institute, Worcester, Massachusetts, p. 73.

[8] Donath, Max, 'Proportional EMG Control for Above Knee Prosthesis', S.M. thesis, MIT, August 1974.

[9] Zlatnik, D., 'Intelligent Controlled Above Knee Prosthesis' in the Proceedings, 4th International Conference on Motion and Vibration Control, MOVIC '98, Zurich, 1998, pp. 1225–32.

[10] Kalanovic, V., 'Feedback Error Learning Neural Network for Trans-femoral Prosthesis', *IEEE Transaction Rehabilitation Engineering*, Vol. 8, no. 1, March 2000.

[11] Grimes, Donald Lee, 'An Active Multimode Above Knee Prosthesis Controller', Ph. D. thesis, MIT D Space Mechanical Engineering, Ph.D., Sc.D. collections, 1979.

[12] OttoBock; home: www.ottobock.com; Endolite Prosthetic Limbs: www.blatchford.co.uk; Victhom Prosthesis: www.victhom.com.

[13] Sonker, Varun, 'Developing Software Suite for Microcontroller-based Above Knee Active Prosthesis', M.Tech. thesis, IIIT-A, 2006; Rai, Lokesh, 'Developing Damping Profile for Prosthetic Legs', M.Tech thesis, IIIT-A, 2006; Agarwal, Amit, 'Study the Effects of Upper Body Motion on Humanoid Walking', M.Tech. thesis, IIIT-A, 2006.

[14] AT90S8535 Microcontroller Datasheet, http://www.nikhef.nl/pub/departments/et/ misc/ at8535/at8535_dat.pdf.

[15] ADC8088 Analog to digital convertor Datasheet http://www.datasheetcatalog.com/ datasheets_pdf/ D/A/C/0/DAC0800.shtml.

[16] CodeVisionAVR C Compiler http://www.hpinfotech.ro/html/download.htm.

[17] AVR Studio 4.12.460; http://www.equinoxtech.com/ products/downloads.asp?details=9.

[18] Fast AVR ver.4.1.3 http://www.ewetel.net/~bamberg.monsees/softeng.html.

14

Information Technology and Education of the Visually Impaired

A. K. Mittal

Regional Director, National Institute for the Visually Handicapped, Regional Centre, 522, Trunk Road, Poonamallee, Chennai-600056 (TN)

Introduction

Endeavours in the fields of educating the disabled, particularly the visually impaired, in our country, have to be viewed in the broader framework of providing and facilitating maximum opportunities for equality, participation and main-stream inclusion. It is towards this end that all our efforts of technology-mobilization, application and refinement as well as technology-transfer must be directed. Indeed, the success of our technological innovations would be directly proportional to the extent to which these contribute towards preparing the disabled learner to cope with the challenges of life effectively and take up the roles of contributing and responsible citizenship. It is, here, that information technology could play a crucial role.

Perspectives

If the twentieth century was an age of science, its successor can rightly be called the age of information technology, which is, of course, an offshoot of scientific developments and innovations. There is no reason why disabled

Disability Rehabilitation Management through ICT, 139–147.

learners should lag in accessing the fruits of the ever-widening applications of information technologies.

Various important developments at the regional and international levels have highlighted the crucial significance of information technology (IT) for suitably empowering persons with disabilities. The UN Economic and Social Commission For Asia and the Pacific (UNESCAP) extended its Decade of the Disabled to 2003–2012. It adopted the landmark Regional Framework for Action to be undertaken for the benefit of persons with disabilities during the extended decade. The Framework identified seven areas for priority action, one of which is 'access to information and communications, including information, communication and assistive technologies.' The goal is to strive towards the creation of an inclusive, barrier-free and rights-based society for persons with disabilities. The UN ESCAP recognizes that ICT can act as a powerful engine of educational advancement and economic growth for disabled persons. It calls upon all member governments to promulgate and enforce laws, policies and programmes to monitor and protect the right of persons with disabilities to information and communication.

Yet another momentous milestone was reached with regard to empowerment of persons with disabilities, when the UN General Assembly adopted the much-awaited Convention on the Rights of Persons with Disabilities in December 2006. Article 4 (G) of the Convention makes it a general obligation of States Parties 'to undertake or promote research and development of, and to promote the availability and use of new technologies, including information and communication technologies, mobility aids, devices and assistive technologies suitable for persons with disabilities, giving priority to technologies at an affordable cost.'

Again, Article 9 of the Convention calls upon States Parties to 'take appropriate measures to ensure to persons with disabilities access, on an equal basis with others, to the physical environment, to transportation, to information and communications, including information and communications technologies and systems, and to other facilities and services open or provided to the public, both in urban and in rural areas.'

Thus, sharing the fruits of information technology is now well-established as a basic right of persons with disabilities.

India has ratified this important UN Convention and, as such, it is only natural and appropriate that information technology initiatives may be directed increasingly towards enabling visually impaired and other disabled persons to overcome the restricting implications of their disabilities and join mainstream society as independent and contributing citizens.

In February 2006 the Indian Government announced a Comprehensive National Policy For Persons With Disabilities. Para 53 (VI) of this Policy envisages initiating research for developing adaptive technologies 'focusing on enhanced personal mobility, verbal/non-verbal communication, design changes in articles of every day usage etc., with a view to develop cost effective, user-friendly and durable aids and appliances with the help of premier technological institutes.' (Government of India, Ministry of Social Justice & Empowerment document No. 3-a/1993-DD.III).

Further, Paragraph 54 of the Policy stipulates: 'Ministry of Science & Technology shall set up Rehabilitation Technology Centre for coordinating and undertaking research and development, testing and certifying technologies, training, etc. Appropriate hardware and software suitable for persons with disabilities to ensure access to information technologies will be developed.'

Thus, the stage is now set for information technology initiatives to be extended suitably and comprehensively to cover the needs of visually impaired and other disabled persons in our country.

The Ministry of Science and Technology proposes to set up a Rehabilitation Technology Centre for helping persons with disabilities through coordinating and undertaking research and development, testing and certifying technologies, training, etc. Appropriate hardware and software suitable for persons with disabilities to ensure access to information technologies will be developed [1].

Technologies Abroad — An Overview

Innovative efforts based on a wide range of IT. applications in Japan, Europe, Australia, Canada, and the US have led to the development of a large number of assistive devices which throw open ever-widening and exciting vistas of opportunities to an increasing number of disabled persons, including the visually impaired.

For instance, access to the printed word has traditionally been the single most challenging area for the visually impaired. Visually disabled learners have had to face obstacles every step of the way, since no solution was ever foolproof. Braille, the best available alternative, was limited in scope because of high production costs and non-ergonomic functionality. Live readers and taped books offered some respite to visually impaired students, but the extent and coverage of material was minuscule. Information technology finally came to the rescue with landmark innovations, such as the Opticon, a device with vibro-tactile output of print letters, the initial version of the Kurzweil reading machine and the Versa Braille, a paperless device, all invented during the 1970s and 80s, The ability of the current adaptive technology to access the Internet and the relative affordability of book-reading machines alongwith extensive magnifying facilities, have, now, made it possible for blind and visually impaired learners, not only to attend and excel at school and college with a high degree of independence but also harness such technological support to stay competitive once suitable employment is procured. Access to computers and the Internet has now become much simpler for the visually disabled user through speech synthessers and screen readers like Job Access With Speech (JAWS), Window Eyes or Window Bridge.

Translation software, such as the Duxbury, facilitate instant conversion from Print to Braille and vice-versa. Computer-aided Braille embossers make it possible to produce material in Braille at an almost incredible speed of 400 to 800 characters a second on paper as well as on zinc plates. These embossers also eliminate the need for Braille-knowing operators and trained Braille proofreaders, at least insofar as Braille production of books in English is concerned.

Technology has also been developed in France, Australia and other countries to facilitate Braille output of the print page being scanned. Systems are now available, which use the six Braille keys for entering texts with output being possible in Braille as well as in print. Additional features on such devices include functions of a talking watch, talking calculator, address/telephone notebook and appointments diary.

A wide range of integrated reading machines are also on the market, making use of OCR technology and providing speech output in English and many other languages; Indian languages however are not yet included. 'Talking books' are available in different models, the simplest being the four-track cassette recorder, which could play back an ordinary C-60 or C-90 cassette with upto four hours of recorded matter. Digital recording techniques are being used for preparing talking books on CD's. This remarkable innovation initiated by Sony Corporation in collaboration with a consortium, is named Daisy.

Then there are talking 'Language Masters', extremely light-weight and easy to operate. The Dictionary section of such devices has as many as 3,00,000 definitions and the thesaurus, 50,0000 synonyms. Also on the market is a highly innovative and useful text retrieval programme, which locates and speaks out or displays in Braille the text and tables from The World Book Encyclopedia and searches for information in the Encyclopedia's 17000 Articles and 1700 Tables.

There is also the well-known Gutenberg Project from United Kingdom and United States, which has compiled an impressive catalogue of electronic books and placed their entire unedited versions on the net. Like all other net resources, these electronic books, too, can be downloaded and accessed using the computer's in-built audio system, supported by software packages of adaptive technology.

Still more is waiting in the wings for the disabled learner. Emerging technologies, such as WAP-enabled internet interfaces can be customized, so that large groups of visually impaired users can listen to relevant web-based materials on a cellular phone, especially, in outdoor settings, such as historical trips.

People with low vision and deaf-blind persons also have a number of IT-application possibilities to avail, particularly, in where large print material for low vision and interpersonal communications for those with the twin losses of sight and hearing is concerned. Online access to leading newspapers and journals is also possible in the West with the exciting opportunities of accessing the articles or news-stories of one's choice.

Access to Distance Education

The dawn of the age of the Internet and the guaranteed ability of the visually disabled user to access it through adaptive technology has brought learning, in general, and distance learning, in particular, to this user's doorstep. Typically, blind students at all levels had to live either inside or within easy reach of educational institutions, such as schools for the blind or colleges and universitieswhich that provide admission to disabled students. Often, this required students to be uprooted from familiar surroundings far from families and friends to unknown towns and cities. Social research has revealed that blind people function best when they are in settings with which they are familiar. Now that many distance education schools and colleges have placed their entire curricula on the Internet in the form of audio lectures, presentations and seminars; blind and visually impaired students, with some supervision and with reasonable access technology support, can attempt to pursue their academic goals both in formal as well as in continuing educational and professional development programmes.

There are, today, a number of full-fledged online universities or training centres in the West, which offer scores of degrees, diplomas and certificates in a wide variety of courses. Many are quite affordable even by Indian Rupee conversion norms. Some even offer learning resources at no cost to the user. Many use audio streaming and wave file download technology to support the printed word. Mentoring is conducted through group and individual chat, e-mail and interactive Web-enabled CBT's. Most issue certificates upon course completion.

The Indian Situation

In sharp contrast to innovations in the West, we in India have mostly been able to utilise, so far, only the simplest, even rudimentary technology, in the development and production of assistive devices in the areas of learning and teaching for the disabled, especially, the visually impaired. It was largely during the last decade that a few institutions of science and technology, willingly undertook work for harnessing information technology for educating persons with disabilities. Yet, barring a very few exceptions, these interventions have not, so far resulted in disabled persons gaining access to really meaningful and cost-effective software packages or other IT-based devices.

Nevertheless, mention must be made, here, of a few pioneering efforts. Organisations in Bangalore, Chennai and Pune have attempted to develop packages to enable visually handicapped persons to work on computers in Indian languages also, though these still need to be extensively tried out and further refined. The prototype of a computerised classroom teaching system for visually impaired children has also been available for some time, although its production on a large scale is still awaited. Data entry through the six Braille keys and a communication device for the deaf-blind are other innovations which still need to be more widely tried out.

Looking Ahead

Adaptive technology has and will continue to offer the challenge of affordability. To partly address this problem in a country like ours, we propose
a community model of computer-applications for the disabled. Such centres
could be established in remote towns and cities. This community centre would
need at least one full-time trainer and a few systems equipped with requisite
adaptive technologies, depending on the size and strength of enrolment. The
trainer should be able to train a cross-section of disabled students in the use
of adaptive technology, Internet access and course curriculum selection. More
work-stations and staff could be added as enrolment increased. These centers
could also act as communicators of distance learning lessons, which might be
prepared by experts in advance.

Some of these centers could also be connected to the National Open School
System and the Indra Gandhi National Open University. Both these organisations have established a web presence, which incorporates a large body of
resource materials to augment traditional distance-learning approaches. The
task of running these centres could be entrusted to reputed voluntary organisations with adequate support from state and central governments.

The captains of our industries also have a vital role to play here. They could
consider sponsoring IT-based solutions for the benefit of disabled learners.
They could also extend necessary support and encouragement towards promoting R&D work in the area. In fact, a proposal was made some time back at
a meeting of the Technology Committee for the Blind under the World Blind
Union, an apex international body in the field of work for the visually impaired.
It was proposed that multinational corporations working in developing countries be persuaded to earmark at least one percent of their profits for supporting
technological and other research activities for persons with disabilities.

The Indian Persons with Disabilities Act of 1995 rightly lays stress through
Sections 28, 42 and 48, on appropriate governments to promote and sponsor
research for designing and developing needed assistive devices, teaching aids,
etc., for disabled children and adults. This should be accessed for research in IT
applications also. The Science and Technology Project in the Mission Mode,
conducted under the Union Ministry of Social Justice and Empowerment,
seeks to fulfill this important role.

Further, in view of our limited resources, work relating to technological innovations needs to be undertaken on the basis of carefully assigned priorities, which should be set strictly as per actual educational needs and the present level of economic development.

Our devices would have to be low cost or heavily subsidized with easy availability of maintenance facilities, wherever necessary. It is heartening to note that the Government of India have recently approved the strengthening of its ADIP Scheme for the Disabled by raising the admissible cost of devices allowed under the Scheme to Rs.25,000. This should help in making IT solutions/devices easily accessible to visually impaired persons. In other cases, the cooperation of the corporate sector may be enlisted to subsidize the cost of such devices to make them affordable for the average disabled user.

Our institutes of technology and other scientific organizations should be suitably motivated to lend their expertise more actively to the development of new devices and packages. Measures like meaningful awards, greater recognition and more liberal funding from various sources may be considered for the purpose.

Conclusion

It is imperative that we follow a three-fold approach of providing all possible encouragement and incentives to our reputed scientists and technical institutions so they can more more actively undertake research and development activities to support persons with visual impairment; of ensuring that suitable and need-based prototypes are developed and the gap between laboratory work and production is eliminated and of facilitating the wider acceptance of access to information technology benefits as a basic right of persons with disabilities. Our immediate and pressing goal has to be technologies and solutions, which are cost-effective and easy to use. Disabled persons' organizations have a crucial role to play in ensuring that their members are able to acquire the devices as per their needs and situation and that the devices under development fully cater to the felt and actual requirements of the users, and not just on those based on the perceptions and ideas of a few well-meaning technical workers/agencies.

Towards this end, we need and solicit the willing and active cooperation of all concerned government bodies, the corporate sector, technological institutions professionals in the field of visual impairment, visually impaired users themselves and their organizations.

References

[1]   http://http://kodakkalshivaprasad.wetpaint.com/page/Disabled+Welfare+Actions

15

Virtual Classroom-based Model for Rehabilitation of Persons with Communication Disorders

Ajish K. Abraham[1], Dr. S. R. Savithri
and Dr. Vijayalaksmi Basavaraj[2,*]

[1]*HOD — Electronics, All India Institute of Speech and Hearing, Naimisham campus
Manasagangothri, Mysore-570006; ajish68@yahoo.co.in*
[2]*Director, All Institute of Speech & Hearing, Manasagangothri, Mysore-570006*

The virtual classroom is a teaching and learning environment which helps to improve access to advanced educational experiences by allowing students and instructors to participate in remote learning communities and to improve the quality and effectiveness of education by using Information and Communication Technology (ICT) to support a collaborative learning process. Unlike satellite-based technology, this medium offers real-time, multi-way audio and visual presence of the professor and the students. The students can see and hear the teacher live, view slides, audio-visuals, participate in interactive sessions with the teacher and other students. Those who miss the class can catch up with the help of video recording. This paper reviews the technical functionality to create a Virtual Classroom to support distance education and evaluates the

*Ajish K. Abraham is Reader and Head, Department of Electronics, All India Institute of Speech and Hearing All India Institute of Speech and Hearing (AIISH), Mysore; Dr. S. R. Savithri is Professor, Department of Speech Language Sciences, (AIISH), Mysore; and Dr. Vijayalaksmi Basavaraj is Director, (AIISH), Mysore.

Disability Rehabilitation Management through ICT, 149–157.

effectiveness of this approach to remote education. Considering the acute shortage of speech and hearing professionals available in the country and the non-availability of instructors at the geographically spread locations, virtual classroom is the only solution for implementing distance learning in this field. The salient feature of the AIISH model is that the complete system is fully automatic and does not require any skilled manpower for operation, either at the central site (AIISH) or at the remote centres.

Introduction

Fundamental to computer-mediated communication systems is the concept of being able to utilize the capabilities of Information and Communication Technology (ICT) to tailor a human communication process to the nature of the application and the nature of the group undertaking this application (Hiltz and Turoff, 1978, 1993; Turoff, 1991).

According to an NSSO Survey (2003), there are about 291 persons with hearing impairment, 94 persons with mental retardation, and 107 persons with speech impairments per 1 lakh persons. The prevalence of hearing impairment is 6.3 per cent as per WHO statistics for South East Asia. The actual number of speech and hearing impaired may be even more than that provided by the NSSO because the survey is conducted by untrained persons and the survey includes persons above the age of four years. Therefore, the prevalence of speech and hearing disorders may be somewhere 0.3 to 6.3 per cent.

Currently there are only 25 institutions in India involved in manpower generation in the area of speech and hearing. These centres bring out only 615 undergraduates and 215 postgraduates in the field of speech and hearing annually. Even if the conservative estimate of NSSO (2003) is taken as the prevalence of speech and hearing problems (592/lakh population), it still means that each speech and hearing professional in India will have to serve about 8,000 persons every year. In reality, a speech and hearing specialist can best cater to the therapeutic needs of not more than 100 persons in a year. This means that a minimum of 50,000 speech and hearing professionals are needed. Thus, there is a vast gap between the resources required and resources available in this field.

Graduate and postgraduate professionals in speech and hearing are involved in the detailed diagnosis and finer aspects of the rehabilitation of people with communication disorders. For early identification and basic rehabilitation of the hearing impaired, what is required is more of hands-on skills than in-depth theoretical knowledge. A diploma course in hearing language and speech (DHLS) would fulfill the need and thus there is a need to start such courses in the regions where they are required to work. Non-availability of instructors even to teach this diploma course in these regions is the major bottleneck and hence the conventional mode can't be adopted to run this programme.

The All India Institute of Speech and Hearing (AIISH), Mysore, one of the premier institutes in the world in the field of speech and hearing, took up this challenge and successfully deployed a novel model of establishing virtual classrooms. The area of speech and hearing being a clinical one, the DHLS programme envisages one-to-one clinical supervision by the staff available in the participating institutions. In the event that staff is not available in the participating institution, AIISH will make arrangements to hire speech and hearing professionals available in the neighbourhood of the participating institutions or even depute its own staff members to run the clinical programme in the initial years. Therefore, the participating institutions should have a minimum of two clinical supervisors to supervise the clinical work of students on a daily basis and to coordinate/initiate other work of the centre.

The institute also prepared the resource material required for this programme. It is felt necessary that resource material should be prepared for diploma students in the local languages. Furthermore, resource material will be made available in printed as well as electronic form.

Objectives

Our objective is not to merely duplicate the characteristics and effectiveness of the face-to-face class. Rather, we can use the powers of the ICT to actually do better than what normally occurs in conventional classrooms. The objectives of our model are:

- Imparting theoretical training by live telecast of lectures conducted at AIISH, Mysore, to four remote centres.
- Creating a virtual classroom at each of the remote centres with a facility for live two-way audio and video interaction between all participants and faculty during the lecture.

- Imparting high quality clinical skills by real-time transmission of demonstration of diagnostic and therapeutic procedures from the clinics of AIISH, Mysore.
- Providing a flexible and easily upgradable technology solution for the participating centres.

Method

Four medical institutions/other organizations willing to participate in this collaborative exercise were identified. Some of these institutions have speech pathologists/audiologists, equipment like audiometer, immittance meter, BERA and OAE (British Education Research Association and otoacoustic emissions) for audiological testing, sound treated rooms, and basic infrastructure like classrooms. More than anything, these institutions have a very good clinical population with various speech, language and hearing disorders that will provide an excellent training ground for students in the area of speech and hearing. Speech and hearing disorders are often associated with other medical problems and hence need a multi-disciplinary approach. These institutions will be ideal places for a multi-disciplinary approach and the patient will get all facilities under one roof.

AIISH uses a Multi Protocol Label Switching (MPLS)-based Virtual Private Network (VPN) to telecast lecture sessions live to four remote centres located at New Delhi, Imphal, Pondicherry and Mumbai. It requires setting up virtual classrooms at these identified remote centres where participants come to attend classes at predefined time slots. The classes are conducted in the central site of AIISH classroom at Mysore. As illustrated in Figure 15.1, the image of the lecturer along with the multimedia presentation or writing on an electronic board is telecast to all remote centres through video conferencing over the MPLS VPN link. It is received by the video conference equipment at the virtual classroom at all centres simultaneously and is projected on a screen through a multimedia projector. The image of the students at the virtual classrooms of each remote centre is received at the central site at AIISH and is displayed on a 32" LCD monitor. Thus the lecturer at AIISH would be able to see the students at all the virtual classrooms along with the students physically sitting in front of him/her at Mysore.

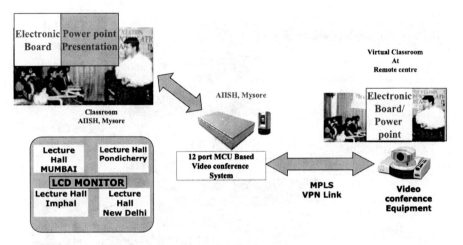

Fig. 15.1

Set up at AIISH Classroom

The equipment schematic of the central site at the classroom of AIISH is shown in Figure 15.2. The image of the lecturer is automatically tracked by the face tracking camera. The voice of the lecturer is captured with a wireless lapel microphone. Thus the lecturer is free to move anywhere in the classroom like in a conventional classroom. The students at the central classroom are captured by two other cameras linked to the digital conference system. Table microphones are provided in each row of the classroom. Whenever a student wants to interact with the teacher, he/she has to press the microphone button and the camera will get automatically focused onto him/her. The lecturer can use the electronic board in the same manner as he uses a conventional board. The scribbling on the board is electronically captured and then coupled to the video conference equipment through a data solution box. The lecturer is free to use PowerPoint presentations in the classroom which is also coupled to the video conference equipment through a data solution box. The lecturer has to toggle between these two inputs at the data solution box so that one of these will go directly to all virtual classrooms along with the image of the lecturer. The video conference equipment is connected to a Multi Conference Unit (MCU), which acts as the nodal control for the entire system. The conference layout at each virtual classroom and also at the central site can

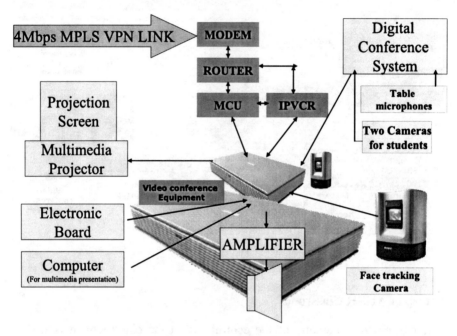

Fig. 15.2 Schematic of Set Up at the Classroom of AIISH, Mysore.

be individually programmed and controlled by the MCU. The system is also linked to an IPVCR (Internet Protocol VCR) which will automatically record all the sessions telecast through the system. The IPVCR can store upto 500 hours of recording. Any remote centre which has missed any of the sessions can replay the missed session by directly dialing the IPVCR from their end. The central site is linked to the four Mbps MPLS VPN network through a router and modem.

Set up at each Remote Centre

The remote centre is linked to the central site through a 1 Mbps MPLS VPN link as shown in Figure 15.3. The data from the link is routed to the video conference equipment through the router and modem located at the virtual classroom. The video conference equipment receives the data (multimedia/electronic board) along with the image of the lecturer from the central site and displays it on a multimedia projection screen. The voice of the lecturer is reproduced by a public address amplifier and speaker connected to the video conference

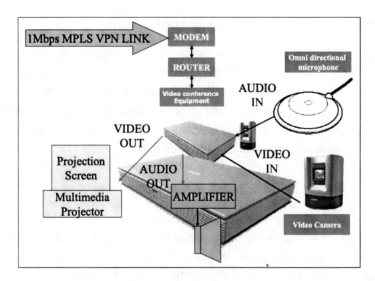

Fig. 15.3 Schematic of virtual classroom at each remote centre.

equipment. The image of the students is captured by a video camera which can pan, tilt and zoom. The camera also has the facility to store six preset image positions. Using this feature, each preset can be adjusted in such a way that it captures the image of the students row-wise. The students can toggle between the preset images depending on the row in which the student asking the question is seated. The voice of the student is captured by an omni-directional microphone kept in front of the classroom. This microphone can be switched 'on' or 'muted' with a remote control so that it is activated only when someone is interacting with the lecturer.

Analysis of Effectiveness of the AIISH Model of Virtual Classroom

It Engenders the Sense of Community Existence among Students

The synchronous mode of communication in our virtual classroom allows immediate feedback on queries and encouragement, which provides the students with a sense of belonging to a community. An instructor can be personally involved with the students during the course of an online chat. This sense of a group existence helps students to identify with fellow learners, thus keeping them glued through the course till its completion.

It Helps Assess Learning Trends among Learners

Similar to a conventional classroom, our virtual classroom provides clues of students' performance, interest and activities inside the class. An analysis of the responses and comments of the participants provides substantial cues to the level of interest and comprehension among learners. For example, the frequency of comments posted by a learner or the type of learner queries often gives an insight into the student's involvement.

It Allows Informative Exchange of Knowledge

An equal and active participation from both the learner and the instructor, and a precise two-way instructive communication between them leads to the fruition of a learning process. Our model of the virtual classroom assists in a meaningful, useful and informative exchange of notes between the instructor-learner, and learner-learner.

It Allows Learning in an Accustomed Ambiance

Learning has always been more effective when it happens in a familiar ambi-ence. The cumulative effect of this results in an augmented, open and active student participation in virtual classroom discussions, which certainly beats that in a traditional one. Simply put, a virtual classroom liberates students from the shackles of geographical distance, discrimination, and inhibitions.

It Provides an Egalitarian Platform for Learners

Students are spared the hassle of dealing with discrimination on the basis of gender, race, or any physical constraints. And even though the instructor leads the discussion, he is recognized by the students only by his screen name. The virtual classroom helps learners lose their fear of teachers — a part of the traditional classroom set-up — although the instructor remains the guiding force of the whole class.

Conclusion

Web-based learning cannot rival conventional classroom training for some very obvious reasons. Since it is mostly delivered through asynchronous communication methods, it provides thin scope for interaction among the learners as well as between the learner and the instructor. It lacks the depth that an instant face-to-face dialogue can provide, because synchronous mode of learning is devoid of instant responses and interaction. In turn, it limits the comprehension of the student, thus reducing his/her interest in learning. On the other hand, our model based on virtual classroom rids web-based training of such maladies. It is essentially a cyber classroom, where the instructor and the learners can converse in real time. Thus, a virtual classroom tries to simulate in every way it can, the learning platform provided by conventional classroom, and is quite successful at that. We have experienced how it genuinely enhances the effect of other learning components in the learning process.

Acknowledgements

We sincerely acknowledge Dr. R.K. Srivastava, Director General of Health Services, and Mr. Deepak Gupta, Additonal Secretary, Ministry of Health and Family Welfare, Government of India, for assigning the programme to AIISH and for all the support. The authors sincerely acknowledge Prof. M. U. Deshpande, IIT Mumbai, Prof. Kannan Modgalia, IIT Mumbai, Mr. T. V. Venkatram, DGM, RTTC, Mysore, and Mr. V. Murali, DGM, Medical Research Foundation, Chennai, for all the guidance given at the design and planning stage.

References

[1] Murray Turoff, 'Designing a Virtual Classroom', International Conference on Computer Assisted Instruction (ICCAI '95), National Chiao Tung University, 7–10 March 1995.
[2] 'Virtual Classroom — A Learning Tool of Prowess', http://elearning-india.com/
[3] Saraswati Krithivasan, Malati Baru and Sridhar Iyer, 'Experiences with an interactive satellite based distance education program', in R. Ferdig et al., eds., *Proceedings of Society for Information Technology & Teacher Education International Conference*, Chesapeake, VA: AACE, 2004.
[4] Saraswathi Krithivasan and Sridhar Iyer, 'To beam or To stream: Satellite-based versus Streaming-based infrastructure for distance education', World Conference on Education and Multimedia (ED-MEDIA), Lugano, Switzerland, June 2004.
[5] Prof. M.U. Deshpande, 'Synchronous "Learning at a Distance" Using Satellite Video Conferencing, IIT Bombay's Experience'.

16

MICOLE: A Multimodal Learning Environment for Visually Impaired and Sighted Children

Erika Tanhua-Piiroinen and Roope Raisamo

The project Multimodal Collaboration Environment for Inclusion of Visually Impaired Children (MICOLE) was a European project aimed at developing a system that could support collaboration between visually impaired children and sighted children within a learning context. The larger aim of the project was to find solutions to improve the inclusion of visually impaired people, not only in school contexts but also later in their working life.

People with sensory disabilities are often integrated in normal school classes but the inclusion does not work well in all cases. Visually impaired pupils can feel like outsiders, especially when doing group work with their sighted partners [1]. The assistive tools they use are designed for a single user, and the sighted participants use mostly visual information to supplement the sense of hearing in collaborative work. For example, in mathematics even in cases when visually impaired pupils were able to do mathematics quite well with their special equipment, sharing mathematical contents with their teachers or peers proved extremely difficult or even impossible [2].

In the MICOLE system the multimodal interaction technology was applied based on the needs and abilities of visually impaired children. Children were involved in the project from the very beginning by taking part in interviews,

Disability Rehabilitation Management through ICT, 159–169.

attending the focus group meetings and participating in usability tests. During the project several prototypes were created and tested, which aimed to improve the collaboration between visually impaired and sighted people.

Prototypes

The learning content of the prototypes varied from geometry to drawing, and included, for example, an application for exploring different angles [3], a space application [4] for exploring different planets of our solar system, an electric circuit application [5] for learning about the components and functions of the circuits, a drawing application [6] for editing and exploring haptic drawings and a multimodal pong game [7], which used a multifaceted audio system with haptics for helping players to coordinate their actions. Prototypes (see Figures 16.1–16.2) were introduced also in MICOLE Deliverables D9, D10 and D15 [8].

Brief Description of the System

The MICOLE system was based on a system architecture that supports collaborative activities utilizing different modalities (visual, audio and haptic). Collaboration is supported, for example, by connecting two haptic devices in the system and allowing pupils to explore and manipulate objects together. Joint manipulation of objects was used especially in geometry applications [9]. People can guide each other using an ordinary computer mouse [10] or using a haptic device as in Figure 16.3 [11]. Two-handed information can be used for giving visually impaired people information about contextual changes in the

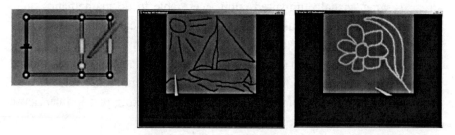

Fig. 16.1 Screenshots of the Haptic User Interfaces of the Electric Circuit (left) and the Drawing Application (middle and right). The Stylus (a cursor) of the Phantom can be seen in all the pictures.

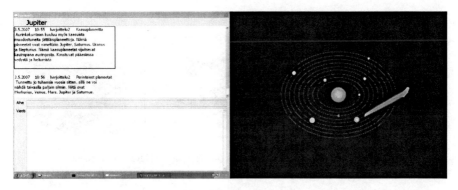

Fig. 16.2 Screenshots of the Solar System Application: The Visual Notebook UI (left) and the Haptic Solar System UI (right).

Fig. 16.3 Sighted and Visually Impaired Children using the Phantom Desktop (right) and the Magellan Space Mouse (left) devices.

collaborative learning environment [11]. A possibility to make notes inside an application was implemented as well [12].

In this project the hardware included several haptic devices like Phantom Omni, Phantom Desktop and Phantom Premium in Figure 16.4 [14], VTPlayer Mouse with two built-in Braille displays [13], Magellan Space Mouse [14] and several audio set-ups in the applications. However, new devices can easily

Fig. 16.4 A Phantom Desktop (left), Phantom Omni (middle) and VTPlayer Mouse (right).

be added to the architecture, which makes the system adaptable for different circumstances and purposes.

The architecture is based on an agent system which allows developers to easily implement multimodal multi-user applications. It is scalable and easily extendable; e.g., new devices can be added to the system just by programming a couple of additional agents that inherit most of their functionality from the base classes. The system consists of several software agents that communicate with each other through a shared bus that is used as a communication channel [15]. The agents send information to the channel in form of ASCII strings, called messages. These messages are available for several agents in the system and a certain agent knows what information is relevant to itself using bindings, which consist of a list of regular expressions linked to callbacks. A more accurate description of the architecture can be found in [16].

Evaluation

The system was evaluated during the project, for example by investigating different collaborative possibilities in pilot studies on early prototypes [17]. Final tests were conducted with five application prototypes: the AHEAD–an Audio-Haptic line Drawing Editor and Explorer, the Solar system application, the Electrical Circuits application, the MAWEN–a mathematical Braille parser and King Pong–an auditory game [18]. The Solar System and the Electric Circuits applications were entirely built on the MICOLE architecture. The Solar system was a part of the earlier space application that was further improved for collaborative use. Testing was done in research laboratories as well as on the field in school situations. The final evaluation was conducted mostly in schools.

The final testing took place in six countries and both teachers and pupils were involved in these tests. Four of the evaluations were conducted in integrated classes with normal lessons and others in separated rooms in schools, except the King Pong testing which occurred outside the school. Altogether 55 visually impaired pupils were involved in the final testing. Testing procedure was similar in every country. The pupils were allowed to do some training with the devices (Phantom for haptics, VTPlayer for tactile cues) before the actual test. The tests based on school tasks (except for King Pong) and the context was also school-like with pedagogic goals to learn something or to accomplish a certain task. After the test the pupils were interviewed. The interview questions were adapted to the different tasks and were not comparable in detail. The test situations were video-recorded and observed with focus on task completion, navigation and exploration problems, behaviour and collaboration [19].

Results

In all the final tests visually impaired and sighted pupils managed to use the applications together, but some challenges and needs for improvements were also evident. An extensive report on final testing can be found; [20] in this paper only results regarding collaboration support are discussed.

One form of collaboration is guiding. Participants of the tests guided or were guided in different ways by the other person in the virtual environments (Figure 16.5). In most cases the guiding took place verbally and was initiated by the sighted party [21]. This guiding occurred very naturally but some navigational challenges were observed. One was correlative of the difference between the virtual context and the way pupils were situated in the physical environment. Saying backwards or forwards meant different directions in real life and in the virtual space.

Force feedback through the Phantom device was also used as a guiding medium [22] and an ordinary computer mouse was used in the Solar system and Electric circuit applications to take the visually impaired person to a certain place in the virtual world. Some usability problems were, however, recognized in Solar System when guiding with the mouse [23]: the speed and the force of the Phantom stylus were too strong when it moved to the target, which surprised the visually impaired children and could make him/her scared. Very strict targets were also troublesome to find even when guided with a mouse.

Fig. 16.5 A Sighted Child (left) uses a Computer Mouse to guide a Visually Impaired Child who is using the Phantom (right).

An important issue regarding group work is that the participants can form a common understanding of the workspace layout and of the actions triggered by other partners. In both the pilot testing as well as in the final tests it was obvious that users were able to achieve a common ground and discuss the task on hand when they explored the same workplace with different senses (see Figure 16.6). However, the design of the multimodal environment affects inclusion during teamwork, which makes it important to carefully plan all the visual, haptic and audio presentations in the environment [24].

Guidelines and Recommendations

During the project we obtained information about user requirements and this information along with the results of the user tests lead us to provide some guidelines for designing multimodal collaborative user interfaces. The guidelines as a whole can be found in Deliverable D14 [25]. In this paper the main guidelines concerning successful group work are presented. The guidelines can be divided into the following themes: context, awareness, guidance, role, initiative and empowerment.

Context

All the members of the group should have equal possibilities to get familiar with the context of group work. The instructions for the common task

Fig. 16.6 A Visually Impaired and a Sighted Child using a Phantom Device together.

should thus be provided to each user in a format she or he can access. It is important that group members have a possibility to check the instructions as many times as they want without having to rely on other group members' interpretation.

The focus should be on creating shared workspaces in order to avoid parallel work processes. Most pieces of information should be in accessible format and if some elements cannot fulfill that demand, it is important to check out if other group members are able to describe these elements properly. Also the possible websites belonging to the environment should be accessible as well.

People should be able to get a quick overview of the shared workplace and to get detailed information of the subsets of it. Every partner should be provided individual means of exploring the workplace and detailed information is needed for locating and interacting with objects of the interface.

For a fluent collaboration it is important to offer group members a common terminology for discussing the manipulation of the environment. This will eliminate misunderstandings and support collaborative error recovery. However, it is recommended to avoid overloading the user with information that is not necessary. In other words, all the information need not be represented equally on different modalities. It is reasonable to decide what is sufficient for the task on hand.

Awareness

When doing group work together people need to keep track both of the communication and the actions of the group. This is important for getting a sense of belonging to the group and for being able to work fluently towards the shared goals. Feedback about the changes that other persons make in an environment is thus needed in order to get a general sense of awareness. Feedback is needed when a person is moving an object, writing a text or drawing an object and it is also needed if a person changes the mode of interaction in the environment. Feedback can be for example a sound when a person grasps an object in order to move it and whenever a person is putting an object down at a new place.

The visual and haptic feedback about the location of the Phantom stylus should be very noticeable when several visually impaired and sighted persons are working in the same interface. This makes the verbal interaction easier and if the representation of the person can be recognized (felt or seen) it can function as a reference point related to the details in the environment for the other group members. People can show direction and intention by pushing the other person, by holding on to the other's proxy or by holding on to the same object.

Guidance

It is reasonable to use haptics to guide disabled users in the workplace and speech to explain what is what in that context. It should, however, be possible for any group member to guide any other member's pointer in the virtual environment. The user of the Phantom device can guide the other Phantom users as well as the mouse user. The mouse user should for one be able to guide the user of the Phantom.

Designers should aim to provide a means of recording and playing back trajectories (see [Figure 16.2]) in the haptic space. This could help in the transmission of shape information and awareness of other users. In real world it is common to facilitate by grabbing and moving the hand of the visually impaired person to the point of interest and in the computer environment this haptic trajectory playback has been shown to be useful for the same purpose. Though the verbal communication is very important in a collaborative working environment for describing simple shapes or scenes, in more complex situations like describing diagrams or complicated shapes the guiding can be supplemented with a trajectory playback. In teacher-learner interaction, this kind of haptic trajectory playback can be useful in situations where a teacher cannot easily find words to describe a certain shape or diagram or when a learner does not understand the verbal description of the teacher.

The haptic trajectory playback should be so customizable that for example the speed of it can be adjusted for different levels of details. Designers should also be aware that large or sharp transitions in a trajectory play-back force can cause the user to lose control of the device. In a collabora-tive environment it is important to include real-time trajectory playback to facilitate verbal communication when describing shapes or trajectories. In other words, the multimodal interaction can be utilized when using trajectory playback. A person can, for example, in a teacher-learner interaction feel a trajectory drawn through playback while listening to the description of the teacher.

Role, Initiative and Empowerment

Collaboration of the users can be supported by taking into account the many ways to present information to the user and using this information for activat-ing role-taking and initiative. Well-presented visual information can activate the sighted user and similarly well-presented haptic information can help the visually impaired person to take an active role in the interaction. When two users share the application, the one which has access to well-formed, content-rich modality with good quality seems to dominate. For example, the user with the haptic device can easily dominate in collaboration if the other modalities are weaker, of poor quality or if their contents are not corresponding.

Providing all the partners engaging in collaborative group work equal possibilities to share the contents and to participate in the workflow facilitates inclusion and that empowers the participants. The MICOLE research project used multimodal interaction technology based on the needs and abilities of visually impaired users. During the project several prototypes were created and tested to improve collaboration between visually impaired and sighted people. Collaboration is supported for example by connecting two haptic devices to the system and allowing pupils to explore and manipulate objects together. People can also guide each other using an ordinary computer mouse or a haptic device. The system has been evaluated during the project and as a result of these evaluations some guidelines have been developed for designing multimodal collaborative user interfaces.

Acknowledgments

We thank all our partners in Europe who participated in the MICOLE (IST-2003-511592 STP) and European Commission for the funding of the project. This work was also supported by the Academy of Finland (decision 114079). Further information from the project can be found at http://micole.cs.uta.fi/.

References

[1] MICOLE D8, 2005, Report on development of collaborative tools –User requirements study and design of collaboration support. http://micole.cs.uta.fi/deliverables_public/deliverables/MICOLE-D8-final.pdf

[2] MICOLE D15, 2008, Deliverable D15: Final evaluation report. http://micole.cs.uta.fi/deliverables_public/deliverables/MICOLE-D15-resubmission-1.1.pdf

[3] E.L. Sallnäs, J. Moll and K. Severinson Eklundh, 'Group Work about Geometrical Concepts among Blind and Sighted Pupils Using Haptic Interfaces', Proceedings of the World Haptics 2007, pp. 330–5.

[4] E. Tanhua-Piiroinen, V. Pasto, et al., 'Supporting Collaboration between Visually Impaired and Sighted Children in a Multimodal Learning Environment', in *Proceedings of the 3rd Haptic and Audio Interaction Workshop*, Jyväskylä, Finland, 15–16 September, Proc. HAID 2008, LNCS 5270, pp. 11–20; VTPlayer. Linux VTPlayer™ Driver (GPL) http://vtplayer.sourceforge.net/

[5] T. Pietrzak, B. Martin, et al., 'The MICOLE Architecture: Multimodal Support for Inclusion of Visually Impaired Children', Proceedings of ICMI 2007, The Ninth International Conference on Multimodal Interfaces, pp. 193–200.

[6] C. Magnusson, K. Rassmus-Gröhn, and H. Eftring, 'A Virtual Haptic-Audio Line Drawing Program', Third International Conference on Enactive Interfaces, 20-1 November, France: Montpellier.

[7] A. Savidis, A. Stamou, and C. Stephanidis, 'An accessible Multimodal Pong Game Space', Proceedings of the 9th ERCIM Workshop 'User Interfaces for All', Bonn, September 2006.

[8] MICOLE D9, 2006, Deliverable D9: Report on results from empirical studies of collaboration in cross-modal interfaces. http://micole.cs.uta.fi/deliverables/deliverables/MICOLE-D9-final-V1_1.pdf; MICOLE D10, 2007, Deliverable D10: Report on the results of collaborative evaluation of the applications. http://micole.cs.uta.fi/deliverables_public/deliverables/MICOLE-D10-final.pdf; MICOLE D15, 2008, Deliverable D15: Final evaluation report. http://micole.cs.uta.fi/deliverables_public/deliverables/MICOLE-D15-resubmission-1.1.pdf

[9] E.L. Sallnäs, K. Bjerstedt-Blom, et al., 'Navigation and Control in Haptic Applications Shared by Blind and Sighted Users', Proceedings of the HAID 2006, Glasgow, UK, 31 August-1 September, LNCS 4129, Berlin/Heidelberg: Springer, pp. 68–80.

[10] E. Tanhua-Piiroinen, V. Pasto, et al., 'Supporting Collaboration between Visually Impaired and Sighted Children in a Multimodal Learning Environment'.

[11] A. Crossan, and S. Brewster, 'MICOLE — Inclusive Interaction for Data Creation, Visualisation and Collaboration', Proceedings of Hands on Haptics Workshop, CHI 2005, 3–4 April 2005.

[12] E. Tanhua-Piiroinen, V. Pasto, et al., 'Supporting Collaboration between Visually Impaired and Sighted Children in a Multimodal Learning Environment'.

[13] VTPlayer, Linux VTPlayer™ Driver (GPL) http://vtplayer.sourceforge.net/.

[14] HP Spacemouse Plus. http://www.dooyoo.co.uk/mice-trackballs/hp-spacemouse-plus/.

[15] T. Pietrzak, B. Martin, et al., 'The MICOLE Architecture: Multimodal Support for Inclusion of Visually Impaired Children', Proceedings of ICMI 2007, The Ninth International Conference on Multimodal Interfaces, pp. 193–200.

[16] Ibid.

[17] E.L. Sallnäs, K. Bjerstedt-Blom, et al., 'Navigation and Control in Haptic Applications Shared by Blind and Sighted Users'; and MICOLE D10, 2007, Deliverable D10: Report on the results of collaborative evaluation of the applications.

[18] MICOLE D15, 2008.

[19] Ibid.

[20] Ibid.

[21] E.L. Sallnäs, K. Bjerstedt-Blom, et al., 'Navigation and Control in Haptic Applications Shared by Blind and Sighted Users'; and E. Tanhua-Piiroinen, V. Pasto, et al., 'Supporting Collaboration between Visually Impaired and Sighted Children in a Multimodal Learning Environment.

[22] E.L. Sallnäs, K. Bjerstedt-Blom, et al., 'Navigation and Control in Haptic Applications Shared by Blind and Sighted Users'.

[23] E. Tanhua-Piiroinen, V. Pasto, et al., 'Supporting Collaboration between Visually Impaired and Sighted Children in a Multimodal Learning Environment.

[24] MICOLE D10, 2007. .

[25] MICOLE D14, 2007. http://micole.cs.uta.fi/deliverables_public/deliverables/MICOLE-D14-final.pdf.

17

Accessibility for All: ICT and Building an Inclusive Society — ONCE's Experience

Miguel Carballeda Piñeiro

Once and the Once Foundation: Digital Inclusion — An Ongoing Aim

The Spanish National Organization of the Blind (ONCE) has, for more than seventy years, been devoted to the tasks for which it was set up: helping the blind and partially sighted people, by means of shared responsibility, to overcome their limitations in everyday life while gradually eradicating social, economic and environmental factors which may cause disadvantages, obstacles or discrimination in their freedom of choice, in reaching their goals, in wellbeing and in personal development.

This means striving to enable the blind and partially sighted to be more independent and autonomous, to achieve full integration and to allow them to participate with the same security and confidence as any other person.

In addition to this task, ONCE acts to foster integration in the labour market, to eliminate architectural barriers and to raise awareness about people with disabilities in society, seeking to gain consideration from society regarding their situation. This latter task is principally carried out through the ONCE Foundation, which plays a key role in normalizing the lives of persons with disabilities in Spain in partnership with other organizations and public bodies.

Disability Rehabilitation Management through ICT, 171–185.

ONCE is able to perform these tasks thanks to its involvement in the public gaming sector by means of a State concession to run a lottery by the name of 'cupón' or coupon and, more recently, other gaming instruments. The income derived from this activity is used to provide for the needs associated with visual impairment of more than 69,000 people in the fields of education, rehabilitation, vocational training, employment, accessibility, culture, new technologies and sport, among others.

Furthermore, through its foundation ONCE contributes to the integration of people with disabilities other than visual impairment in Spain, amounting to approximately 4 million Spanish citizens. Accordingly, ONCE and the ONCE Foundation provide employment to 118,000 people of whom 78 per cent are men and women with some type of disability.

Looking abroad, and mostly through the ONCE Foundation for Solidarity with Blind People in Latin America (FOAL), ONCE also develops cooperation programmes aimed at improving the life quality of the blind in different parts of the world.

Finally and in terms of the European Union, ONCE is traditionally one of the most active social organizations in promoting programmes and policies which have brought positive results for the disability sector. In this field ONCE works in partnership with public institutions and, most importantly, with European bodies in the social sector to which it belongs.

Activities Aimed at the Blind and Partially Sighted People: The Users

In the current setting as regards ICT and the information society, the development of persons with visual impairment in terms of training, employment and social aspects is determined largely by the skills they possess in modern technologies. However, there are currently technical devices in the market that cannot be easily used by everyone. In many cases they must be adapted to ensure they do not become an added disadvantage or bring about further segregation of certain citizens. This situation particularly affects people who are blind or partially sighted given the visual content used in these technologies. On the other hand, in general terms the application of new technologies has been and continues to be a constant source of solutions in terms of autonomy and well-being for this sector of society in different areas such as daily living, mobility, education, employment, free time, culture and others.

To bring about access to and the use and application of such devices and assistance, ONCE works diligently through its network of regional offices and its Centre for Research, Development and Application of Assistive Devices for the Blind (CIDAT).

ONCE's commitment can be witnessed throughout its existence, making a concerted effort in the fields of researching, developing and producing material and technologies, mainly through CIDAT — its specialized centre in this field — and devoting considerable human and financial resources to ensure visually impaired people have the best possible means to be able to join and take part in the information society.

The tangible outcomes of these efforts can be seen not only in the products developed, their dissemination and training blind and partially sighted people to use new technologies, but also in the significant amounts assigned for partnerships aimed at developing communication protocols, website accessibility standards and associated tasks.

CIDAT is a top-level centre with highly-qualified professionals and specialized technical skills. Its mission is to provide ONCE members with the assistive devices they need for personal development in the different environments in which they live their lives, including employment, training, home and socio-cultural settings. While its principal goal is to offer top-quality services to ONCE members, it is also a service provider for domestic and international clients, both individuals and businesses, who approach with requests for services, which include support and technical advisory services, equipment repair and the sale of goods from its range of products.

In addition to this, CIDAT is a centre for the following activities:

- Analysis, assessment and approval of prototypes, devices and programmes with advice on the specific adaptations required to enable them to be used by blind and partially sighted people.
- Designing and developing its own range of accessible devices and products.
- Providing advisory services to businesses and authorities interested in internet accessibility for people with visual impairment.
- Above all, developing research and development projects in the field of accessibility in partnership with other institutions, companies and research centres in the European Union. Among the latest breakthroughs we can find talking mobile telephones, speech

synthesis in household appliances, digital notebooks for blind students, accessibility in virtual reality environments and an ongoing project involving applying GPS technologies in mobility aids for the blind.

Services related to technology are delivered to all users through ONCE's regional structure in the whole of Spain, thus enabling training and assessment to be tailored to each individual through the Communication and Access Division in ONCE's thirty-three Social Services Departments and five Educational Resource Centres. This network of thirty-eight centres also carries out the dissemination and distribution of products in the corresponding outlets/display areas and organizes new product launches.

The division provides members with a number of services. First, the user is given orientation and advice on the assistive technical aids and devices that best suit his or her visual impairment (blindness or severe partial sight), interests and needs. Thorough training is subsequently given to those who want or need it for educational, employment or leisure purposes. Training is adapted to the individual and includes ongoing training to update skills on new versions of tools that may be released.

The average number of people using this service in the last five years–1,200 annually–plus the fact that demand is growing demonstrates the interest and needs persons with severe visual impairment have in using such technologies in their daily life.

The main assistive devices for which training courses are offered are as follows:

- Braille and speaking devices for on-screen information and computer keyboards (Braille displays and text to speech)
- Braille and talking calculators
- Note-takers and electronic agendas
- Adapted software enabling blind people to work on Windows (screen readers for Windows)
- Adapted magnification software for people with residual sight (screen magnifiers)

In addition to using the regional infrastructure outlined above, CIDAT has a special information and assessment telephone hotline and online e-mail service which provides information on its products' features and uses and extends customer care when problems are encountered during the installation or set-up process. It also has a website aimed especially at disseminating news on assistive devices and new products, offering accessibility software for download, swapping material with other users and browsing CIDAT's product catalogue.

In 2004 the centre was granted quality certification by the British Standards Institution, bringing it close to excellence in service provision. This certification proves that all the processes undertaken to perform the different tasks in different areas at the centre are in compliance with the stringent ISO 9001:2000 standards. ONCE is delighted to have received such recognition, and it intends to keep up the good work and continue providing the best standards of service to satisfy its users.

Tifloinnova is an outstanding activity in terms of spreading the message regarding technological breakthroughs to CIDAT's members. *Tifloinnova* is a fair hosted by CIDAT which was held for the second time in 2008. Through exhibitions, conferences, workshops, demonstrations and practical presentations, it aims to send visitors the message that assistive devices for the blind are everyday items, easy to use and close to reality.

The social and commercial impact of events such as *Tifloinnova* is evident from the interest shown by the forty-six providers from sixteen countries that took part in the event in 2008, when 3,000 visitors attended and wide media coverage was given. *Tifloinnova* is thus emerging as a dynamic setting where developers, manufacturers, producers and distributors working in software, hardware and assistive devices for the blind can meet end users and professionals involved in education, employment, rehabilitation, free-time activities and daily living.

While knowing what technological assistance is available and how to use it is vital for digital inclusion for people with visual impairment, it is equally important to have access to the assistance and be able to use it in two key areas — training and employment, and above all if we bear in mind that this type of assistance is expensive. This is where public authorities must draw up clear and decisive strategies to make up for the negative consequences the high cost could imply for individuals.

One of ONCE's key social goals is to ensure equal opportunities for its members both in education/training and in their access, retention and improvement in terms of their jobs. That is why the employment support service focuses on studying, describing and adjusting tasks, procedures, physical environment and tools used by members in their workplace to take into account their visual impairment and the repercussions it has both in communication and in access to information.

By means of a detailed study on the interaction of the factors mentioned above, a specialist (an instructor in assistive devices) determines what accommodations, aid and physical or logical specialized assistive devices are needed to make a position accessible in the most ergonomic, efficient and competitive way possible.

With regard to this, for some years now ONCE has made a strong commitment to include all blind and partially sighted employees in the digital environment in its activities and engage them with its computer software, establishing accessibility and universal design standards when developing systems and delivering both the assistive devices needed and the training required to use those devices.

This approach includes accessible technologies in fields such as bar-code readers to make the job easier for all those who are selling lottery tickets on a daily basis and those who need to check and confirm the veracity and accuracy of winning tickets and their date.

Members who are employed in companies and bodies outside the organization are provided with the accommodation they need to perform their tasks. This is given as a free loan to the individual, and maintenance is also provided at no cost either to the user or the business.

A similar method is used to support students, but in this case it is the teacher who is involved along with the instructor in assistive devices. Material from the mainstream market or assistive devices and accessible software are provided and loaned to students who are blind or severely/partially sighted as part of curricular adaptation to ensure they gain access. Accommodation facilities are provided at the place of study in accordance with the criteria regarding curricular needs and suitability of the material at each stage of the educational process. The purpose is to allow students with disabilities to use digital technologies in order to enjoy more possibilities, have easier access to information and teaching material and resources, and enable them to enjoy a wider range of options in terms of training and employment and, consequently, social expectations.

Given the importance of information provided in visual formats and the growing interest among the general public in taking part in cultural and recreational activities, the AUDESC (or audio description) system is particularly important. This technology is provided to boost access to people with visual impairment to multimedia content such as films, plays, television programmes, and so on. Using different technologies–video mastering, infrared transmitters from a booth and a small earpiece–to adapt and provide oral description of pictures, blind people can not only follow what is happening but also be given information about the setting, clothing, landscape, and gestures and attitudes of the characters.

Although most of ONCE's activities in this field have focussed on audio-describing films and theatre plays performed by professional theatre companies in partnership with public authorities, there has also been some activity recently in audio-describing television broadcasts.

ICT is being used in ever more important fields. In addition to those already cited (cinema, theatre and television), the following are worthy of mention:

- Training and teaching videos are widely used in businesses, schools and universities.
- AUDESC is useful for giving information and describing facilities at visitors' centres in national and theme parks.
- AUDESC is used in exhibitions and museums to describe projected slides, layouts, videos, and so on.

Making the innovative experience that AUDESC offers widely available to people is a necessity that requires concerted support from public authorities and suitable legislation to back its implementation.

One recent initiative that aims to facilitate access to information by blind and partially sighted people and take full advantage of the possibilities new technologies offer is the 'Members' Only Club', a ONCE portal for the exclusive use of its members where they can find information of general interest about the organization; agreements reached by the organization's General Council; ONCE's constitution and by- laws, and so on. The club also offers additional content such as:

- Information on adapted cultural resources (accessible cultural centres, historical and natural sites), a catalogue of audio-described

audiovisual material and a 'what's on' section offering information on shows and plays in the member's home town where the
AUDESC system is used to ensure accessibility.
* News about assistive technologies for the blind, CIDAT's range of
products and information on how to purchase them.
* A monthly agenda of activities and events organized by ONCE in
all of its premises.
* A section designed specifically for older people where they can find
information about ONCE's activities (cultural excursions, physical
exercise, workshops, etc.), clubs for older people, holidays and
volunteering.
* News related to visual impairment or the disability sector in general
which is of interest to members.

The Members' Only Club offers many other very useful services including:

* Digital library: Providing online access to books (novels, poetry,
plays, guides and manuals) produced by ONCE's Library Services
(SBO) in TLO and Daisy format. The former are works produced
in Braille that can be printed on paper or read on a computer using
the TLO programme which has been developed by ONCE and is
available in the 'tools' section of the website. Daisy books can
be played on any of the devices placed at the disposal of users
by CIDAT, or on a computer using either TPB Reader or Daisy
Player software, which is also available for download in the 'tools'
section.
* Employment and training: This section aims to provide visitors with useful resources such as a database of employment
opportunities, information on training (courses, online computer
and language courses) and other employment-related tools. The
employment support service provides the Employment Opportunities Data Base, a new tool to assist users in the job seeking process.
It offers immediate and easy-to-use access to information on current job offers in the ONCE group and outside the organization,
thus giving members the opportunity to apply for any of the positions available and take part in the selection process. In doing so,
the member can also receive orientation from multi-professional
teams based at the ONCE centres mentioned previously.

Support for R+D in Accessible Technologies

Over the last decade ONCE has strongly backed efforts to foster coordinated action in support of initiatives, projects and actions aimed at encouraging developments and technological breakthroughs to modify the environment while taking into account the needs of the visually impaired. In terms of the disability sector, scientific research and technological innovation are very specific tools in the inclusion and social cohesion policies.

Fostering the implementation of the 'design for all' principle has emerged as one of the preferred strategies in the field of technology supporting persons with disabilities. This is set out in the framework of the European project INCLUDE, which aims to design services and products that can be used by the largest possible number of people, taking into account the diversity of human skills and not merely the standard or average skills, without the need for adjustments or specialized design. A second strategy which complements the first is to design products or services specifically when the reduction in skills or ability is such that mainstream products cannot be used, even if these have been designed to take into account the lowest levels of skills.

To date actions have focused on the second strategy because they are more complex and require financial backing. In both cases, however, the goal is to bridge the gap, or at least reduce it, between the functional capacity of the individual and the capacity needed to operate the device or technology, ultimately in order to remove limits to that person's effective participation in society. Whether one strategy or the other is employed will depend on the adjustments to the available technologies for the user.

The reaction from those authorities, organs and bodies that have proven to be most responsive and sensitive to social demands has taken the shape of actions in favour of inclusion such as the 'e-Europe' initiative, the drawing up of guidelines for public administrators in the European PROMISE project, the development of the ISO 9999 international standard related to classifying assistive products, or setting up a Network of Centres of Excellence in Universal Design with the support of the European Commission. To all these efforts we can add ONCE's contribution, aimed at combining and balancing consideration for the needs of its members with the encouragement needed to boost groundbreaking research. In this respect, it is worth mentioning the work done by CIDAT, the ACCEDO Group, the ONCE Foundation and the ONCE International R+D Awards.

Without doubt, one of the outstanding trends in the information society is towards dynamic instruments that generate new contexts for learning, particularly in schools and universities. This involves developing platforms for education which include teaching and learning resources with suitable content for the different stages and levels of teaching in the implementation of digital technologies in the classroom.

A brief study of internet-based resources which are representative of this kind of platform shows that many of them are inaccessible to the technological and access tools used at present by persons with visual impairment. This is the case, above all, if we bear in mind the fact that accessibility standards have focused on internet surfing and have tended to ignore programmes for teaching. It is worth stressing that adjustments to screens, graphics, tables, videos and other elements involves, in many cases, more than merely adding descriptive tags or writing alternative text to provide an explanation.

One of the most outstanding initiatives in the European context in terms of promoting digital inclusion in the classroom is the research carried out jointly by ONCE and Hewlett-Packard on the use of Tablet PCs which include tools for access and adapted embossed material and worksheets in order to assess their utility and accessibility depending on the age, school year and degree of disability.

To overcome the current problems and avoid new forms of exclusion in the school setting, it will be necessary to promote joint and coordinated actions with education authorities from both the public and private sectors and research teams. These initiatives must aim to ensure the highest possible level of accessibility when using digital resources and content in the curriculum, thus enabling students with visual impairment to perform as well as possible in the classroom.

In the field of partnerships in digital inclusion between ONCE and the education authorities, the work carried out by the ONCE Educational Content Accessibility working group is crucial. This is a multi-professional team composed of specialists from ONCE and the ONCE Foundation (teachers, assistive device instructors and technical staff working in production) whose main objective is to advocate for the creation of schools that are completely accessible to people with visual impairment using the breakthroughs new technologies bring.

Among the results achieved, ACCEDO's research activities have helped to draw up guidelines to act as a reference for all professionals involved in designing and developing such platforms and environments in the field of education, making them accessible to and usable by students with all types of visual impairments. ACCEDO also advises and works with public bodies and private sector companies (in the computer and education sectors, and publishers) to develop accessible software and hardware.

ACCEDO's work has spread beyond Spain's borders and has been recognized by the education community in Europe, as witnessed by the prestigious Fifth Handinnov Europe 2008 Award it received. The aim of this award is to reward innovative projects carried out in any European Union member country on behalf of young people with disabilities to aid them in their educational, professional or social lives. This is in line with the objectives set out in the recently-adopted UN Convention on the Rights of Persons with Disabilities.

In addition to the initiatives described above, the ONCE International R+D Awards in new technologies for the blind and partially sighted, organized every two years, is yet another example of the efforts made by ONCE to foster technological developments that contribute to facilitating the autonomy and inclusion of people who are blind or partially sighted. The awards' aim is to contribute to developing top-quality projects with proven feasibility which may result in significant breakthroughs.

As a result of the experience gained from R+D projects carried out by ONCE or in partnership with other organizations, and from an in-depth study undertaken by experts in ICT and assistive solutions for the blind and partially sighted, engineering, artificial intelligence, computing, telecommunications, microtechnology and nanotechnology have been identified as priorities for research in terms of their implementation in the following fields:

- Means of facilitating access to information in digital formats, computers and telematic networks; software for access to information; accessibility of free and open source software (Linux and software designed for Linux); hardware devices for displaying embossed information.
- Access to paper-based information; information conversion and handling software; optical character recognition software; text magnification devices; and so on.

- Braillers and associated software; innovative systems for Braille printing and embossing; Braille conversion software; drivers for printers.
- Educational and recreational software including games and software to help teach and learn subjects such as mathematics, physics, music, etc.
- Systems enabling access to PDAs; software to widen accessibility to mobile telephony environments; new add-on software solutions to existing programmes.
- Access to home automation and electrical appliances; systems and software enabling access to home electrical appliances; developing accessible domotic systems, and so on.
- Public transport: Systems providing access to information at stops and stations; technologies that can be applied in vehicles to give informationin accessible formats.
- Audiovisual field: New technologies in the field of digital television or accessibility to DVDs.

The goal, in short, is to promote any development in the field of science, technology and innovation which will foster accessibility to goods and/or services and contribute to the integration of persons with visual impairment into society by bringing about improvements in their training, mobility, home and professional lives, relationships and communication.

ONCE, through its foundation and in solidarity with the disability sector, also promotes R+D activities to develop solutions that guarantee access to the information society for people with other types of disabilities.

One of these solutions is the INREDIS project (relation interfaces between the environment and persons with disabilities). The project is run by a consortium of businesses led by *Technosite*, a ONCE Foundation technology company. A range of large corporations, finance companies, technology businesses, public bodies and centres for technological research are engaged in this project, thus providing a global perspective of the market, enabling standards transfer among sectors and generating added value opportunities for specialized businesses offering services.

INREDIS is a new approach to applying technologies in digital inclusion. Interoperable, accessible and multi-device are the keywords that best define this project and will shape the technological products of the near future. Research areas include security devices, potential channels for interaction, communication protocols and system interoperability and their application in several disability-related fields, and the inclusion of users with disabilities in various aspects of the information society, including domotics, building automation, mobile telephony, urban and local mobility, shopping information, banking, digital television and other fields of interest.

The scope of these technological outcomes will have a significant social impact worldwide as they will be important breakthroughs in the field of accessibility for persons with disabilities and will bring about noticeable improvements in their life quality.

Challenges for an Inclusive Information Society

During the last decade, political initiatives have generated significant and important social progress based on a new 'social model' of disability that holds people with disabilities to be citizens with full rights. This is the result of the receptiveness of public powers, growing social awareness among European citizens and, to a large extent, the efforts of the disability movement to which ONCE belongs. The new model is based not on the limitations of the individual, but on his or her capabilities, on the paradigms of personal autonomy, life quality and, above all, on the understanding that the ultimate objective is the elimination of physical and social barriers.

The impact of ICT on our society and its development alters the way in which we citizens conduct our relations and affairs, access services and information and, in general terms, communicate.

As a result, access to information and communications is an essential prerequisite for anyone who wishes to take part in society. To enable persons with disabilities to exercise their rights actively, participate in society and engage in decisions affecting them, it is vital that all ICT-based goods, products and services are accessible to them. As this is not always the case, accessibility has become one of the key challenges our society faces.

To bring about an inclusive information society in which people with disabilities in general, and visually impaired people in particular, are able to participate on equal terms and as citizens with full rights, it will be necessary to:

- Improve digital accessibility (e-Accessibility) and usability in ICT tools and services;
- Foster digital literacy and skills;
- Promote the dissemination of and access to assistive devices that enable participation in digital environments, products and services (mobile telephony, digital television, online services, and so on);
- Bridge the potential digital gaps based on geography, age, gender and disability (blindness, partial sight, hearing loss, intellectual disability, physical disability, and so on);
- Boost innovation in and roll out of accessible electronics.

Electronic accessibility constitutes a key element in the European digital inclusion policies. In the wider context, ICT falls within the scope of the implementation of the draft directive on equal treatment, which makes reference to access to publicly-available goods and services and their supply. Member states must also meet the obligations laid down in the UN Convention on the Rights of Persons with Disabilities with regard to the accessibility of ICT goods and services.

In spite of the benefits and the attention given to the matter in the political field, progress in electronic accessibility continues to be unsatisfactory. There remain numerous and striking examples of shortfalls in accessibility.

Consequently, and as we move forward and develop the *eEurope Initiative*, the *i2010 initiative* and the *Riga Ministerial Declaration on an Inclusive Information Society*, it is vital to adopt a common and coherent European approach and put in place public actions in the fields of legislation and financing to favour the 'information society for all', in which disability-related issues must be considered priority, so that the information society is one in which all citizens partake, that is to say an information society that fosters integration and social cohesion, not a new cause of exclusion.

In this respect, ONCE operates from the two perspectives outlined above: on the one hand, it provides blind and partially sighted people with resources that boost their personal autonomy in the environments in which they participate; and, on the other hand, it plays a significant role in the disability movement and collaborates with public powers and private-sector initiatives to promote social integration and accessibility.

Although it is necessary to guarantee that the instruments available in the present legal framework are effectively implemented, it is worth underlining that legislation is not the only available tool at the European, national and regional levels in the fight against discrimination and social exclusion. In practical terms and to promote non-discrimination and inclusion, a wide range of political and economic instruments must also be fully utilized.

In short, to make sure ICT is an opportunity and a unique tool for the inclusion and normalization of persons with disabilities, it is essential to have commitment and concerted action from the various stakeholders involved: public authorities, technological solution and digital service providers and the disability movement to which ONCE belongs. Such concerted action will result in full citizenship for people with disabilities in the information society.

Part IV

Assistive Technology for Differently Abled Persons — the road ahead

Part 4

18

ICT and Rehabilitation Engineering: Changing Digits

Col Dr PD Poonekar[1] and Col Dr PK Gupta[2]

[1]151 BH, C/o 56 APO; ppoonekar@hotmail.com; pdpoonekar2006@yahoo.com
[2]Training Officer, AFMC, Pune

Introduction

'Disabled' persons, now more appropriately described as 'challenged' or 'differently abled' still constitute a large under-served population. Their rehabilitation involves inputs from various fields. However, welfare and rehab programmes have failed to reach the majority of such people. The basis challenges here essentially revolve around poverty, illiteracy, social insensitivity at different parameters and lack of adequate administrative support. Illiteracy and poverty are closely linked and take an added dimension when it is children with disability being dealt with. For various reasons, the policies for the disabled have mainly remained on paper and not been implemented. This often forces the differently abled persons to seek legal resort on already existing executive policies. Information Communication Technology (ICT) is a great boon at this stage, not only for education and training but also provisioning 'rehabilitation engineering' to provide customized end-user devices and their subsequent follow-up and monitoring.

Disability Rehabilitation Management through ICT, 189–194.

Impact of Basic Challenges on Rehabilitation Engineering

The differently abled cannot be rehabilitated in isolation; rather rehabilitation has to be integrated into the existing healthcare milieu in an 'extended' or complementary form. The gamut of the existing healthcare setup in India involves some interesting parameters. As the second-most populated nation in the world, India has only 2.4 per cent of the world's land area. Pertinent data from the 'Economic Research Survey' (1990) shows that the poorer families — 40 per cent — spend an average Rs 157 per illness when receiving care from government doctors and Rs 131 when purchasing care from private doctors. The wealthier 60 per cent paid less to government doctors (Rs 137) and more to private doctors (Rs 215). Exact figures today are difficult to compute due to the dynamics of the progressive presence of many large health insurance players. Present statistics point to 262 recognized medical colleges in India producing 30,922 doctors per year. Most of these colleges are located in urban areas where less than 30 per cent of the population resides. Ironically, 75 per cent of health services here are currently provided by the private sector. The 73 per cent who live in rural areas have no option but government-run Primary Healthcare Centers (PHC) where doctors are often unavailable or altogether absent. Records show that basic specialists' positions to the tune of 55 per cent are vacant in PHCs. Out of the 'listed' 6,83,82 'rural working' doctors (as of 2006) only 1 per cent work in rural areas. Legislative action is at hand to better the situation but with little effect. Thus, the differently abled population calls for networking. In this rather grossly disproportionate healthcare and rehab milieu, ICT can prove to be an effective tool, not only as a potential solution for addressing policies and strategies, but also to optimize and augment the much-needed rehabilitation action, even at grassroots level.

Differently Abled Persons Groupings

Conventionally, the differently abled are grouped under locomotor/orthopaedic disability, visual impairment, speech and hearing disability and mental retardation. More recently, the geriatric group has also been included as a challenged segment. At times an individual may have multiple disabilities and needs to be addressed in a special manner. It is estimated that about 10 per cent of the population suffers from one disability or the other.

Rehabilitation Engineering

Rehabilitation engineering aims at providing devices to facilitate the differently abled not only to overcome their physical disabilities but also help them to become a productive members of society, thereby integrating into it and enjoying a better quality of life.

Rehabilitation engineering for the visually impaired envelopes devices and training related to the Braille system. New innovations include recent inputs on keyboard modifications; more recently 'photosensitive micro-chip systems' have been developed as substitute for retina function.

Devices for the hearing-impaired have received a boon in the form of the micro-processor used as a 'cochlear implant', which converts sound waves into nerve impulses that are linked with normal brain activity.

Locomotor disability constitutes the major bulk of the challenged population. About one million amputees require prosthesis (artificial limbs) and about 1.5 million require wheelchairs/tricycles as part of short- or long-term goals of rehabilitation. With energetic implementation of the 'National Leprosy Eradication Programme', leprosy cases have drastically reduced from 4 million in 1981 to about one million in 2007. Recovered leprosy patients often require prosthesis or orthoses (braces), depending on each deformity. Hence, the major rehabilitation engineering falls under the group of prosthesis and orthoses, i.e., P & O, and related mobility aids.

Rehab Engineering for Prothesis and Orthosis (P & O)

Artificial limbs are provided by 'prosthetists' and braces by 'orthotists' — the programme that trains a specialist in both disciplines leads to a Diploma or a BSc. Such courses are conducted at Mumbai, Kolkata, Cuttack, Bangalore, Patna, Delhi and Karigari. Experts are seriously thinking of introducing a MSc course in the near future in at least one institution and to regularly hold 'bridge courses and related examinations' under the banner of the International Society for Prosthetics and Orthotics (ISPO) for international accreditation. There has been an increase in qualified persons moving abroad to greener pastures in the last few years, but it is estimated that about 600 P & O practitioners and around 800 technicians (with hands-on job training) provide for the locomotor challenged group in India. ICT has helped standardize and bridge these training programmes to a significant extent and bring up P & O end-products to an internationally comparable standard. Wheelchair/tricycle provisioning In India is closely linked with the bicycle industry as the basic materials are a common denominator. This is unlike the scenario in developed countries where wheelchair/tricycle constitutes an elite and independent industry per se.

ICT and Rehab Tools

As the orthopaedically challenged constitute the largest differently abled group, focusing on the impact of ICT and rehabilitation engineering issues would stand as a good paradigm and corollaries drawn herein for other groups. Irrespective of the individual-in-need, the prosthesis or orthosis needs to meet a common standard. It should be of sound bio-medical design; be light, strong, comfortable, durable economical and easy to manufacture; should have a snug fit; should be repairable at multiple places as and when required and should meet regional cultural and traditional values. Additionally, in the last decade there has been growing emphasis on the use of recyclable material and eco-friendly technology. With the last President of India, Dr APJ Abdual Kalam taking a special interest in material sciences, 'space-age materials', such as carbon-fibre composites have become easily available for fabrication of not only P & O items but also for special wheelchairs. These composite materials are gaining popularity due to their weight to strength ratio.

Even the famous 'Jaipur foot' and 'Jaipur limb', initially made from aluminum shank have been gradually replaced with HDPE and PP materials. Plastics in various grades of elasticity, flexibility and compressibility have virtually replaced fabrics, metals and wood. The greater impetus comes from easily available Western products which are converted to appropriate customized devices with either indigenous or hybrid technology — a kind of 'reverse technology'.

Basic Rehab Engineering: P & O Processing

P & O devices are often customized and made by sequential steps consisting of first taking a 'negative' Plaster of Paris (POP) cast of the limb or torso which needs to be provided with the device, then a thick POP suspension being poured in the negative cast to obtain a 'positive'. This in turn is modified by special carving tools to give a modified POP mould. The modified cast after drying is used to make the 'socket' for the prosthesis and the orthoses.

The socket for the prosthesis is then linked with the terminal prosthetic foot for a leg amputee or with intervening shank and joint of prosthetic hand for the upper limb amputee, depending on the level of amputation. The spectrum of socket materials, prosthetic feet, hands, joints and skin forms a large choice of alternatives and a huge window in which to apply rehab engineering. The joints may range from a simple 'pendulum-based' system to systems that incorporates hydraulics; pneumatic, electronic or micro-chipped mechanisms; or the more recently introduced 'myo-electrical impulse'-based devices.

Apart from various government agencies, many NGOs are active through the community based rehabilitation (CBR) programme. ICT has been very useful for not only link-ups and smooth monitoring but also for multifaceted up-gradation of the engineering end products. Fusion of CBR and PHC personnel has seen a healthy evolution of 'multi-purpose rehabilitation workers' (MRW); here ICT is a major interfacing instrument. The biggest potential asset e-rehabilitation which may be developed to the extent that details of limb and deformity can be transferred to fabricate the customized device at a remote place and the device can be speedily sent to the consumer for optimal fit and function. The cosmetic element can be addressed by the use of CAD–CAM system.

Computerized Aided Design and Computerized Aided Manufacture (CAD–CAM)

Computerized-aided design and computerized-aided manufacture (CAD–CAM) involve use of digitized information on an engineering scale by using ultrasound/IR/laser/Rf (radio-frequency) scanner. Recently hand-held sensing devices have been available and so can be interfaced even at a 'rural rehab-engineering' platform. The digitized input is transferred via modem networking (landline/wi-fi/satellite communication process) to a central/zonal 'manufacturing station' for the making of the final device. The digitized information is used by a Computer Numeric Controlmachine at the manufacturing station to make a modified replica of the limb or torso over which the end customized prosthetic socket or orthosis is manufactured by a vacuum or direct plastic molding process. The final usable device can be sent (by road/rail/plane) to the relevant differently abled individual to suit his or her bio-mechanical requirement. Thus, the routine POP process which may usually be spread over a few weeks is telescoped to a few hours, and customized device based on sound bio-mechanic principles fabricated in a remote place without the differently abled having to move out of place. This is analogous to a document being sent from one place to another via digitization and received as a printout without the document really moving at all. Further consultation can be conducted by transfer of still images or time-related images as video formats/real-time transmission for finer adjustments (or added training) as per requirement — the entire process akin to video-conferencing.

Future in Rehab Engineering

The last two decades have seen considerable growth in rehab engineering due to inputs from conventional engineering platforms, such as mechanics, electrical systems, hydraulics, pneumatic electronics, material sciences, polymers and micro-chip systems. Future vistas in rehab engineering will open with principles based on nano-technology and neural-transmission. Considerable research is being carried out to harvest cellular and intra-cellular engineered components for rehab. Progress is being made in engineering cellular growth as an option for spinal cord regeneration potential in paraplegics. Interfacing with 'robotic engineering', neural-prosthesis are working towards an experimental stage being activated by mere thought processes or brain activity. The realization of this will result in a paradigm shift in the way prosthesis or robots are made and the differently abled will be truly rehabilitated into the bionic age. Genetic engineeringholds a different promise via genome modification/modulation. Such an approach would control the basic DNA–RNA based enzymatic processes. All these newer inputs are heavily reliant on ICT in one way or the other. Use of microchip-systems for neural continuity and synapses may seem a fantasy right now, but is a fantasy that sits on the R&D anvil today. Thus, the last word in ICT and rehab engineering has yet to be spoken and the discipline is at an ever-expanding horizon limited only by imagination.

19

ICT and Families of the Differently Abled

Kiran Rao

National Institute of Mental Health And Neuro Sciences (NIMHANS), Hosur Road, Banglore-560029

A majority (about 60 per cent) of persons with severe mental illness live with their families [1]. As a result, it is the family that is involved in most aspects of care. In India, given the lack of alternate systems of institutional care, the number is even greater.

The family plays an important role in the decisions regarding engagement in treatment, supervision of medication, providing day-to-day care and emotional support to the individual [2].

The demands of caring for a person with severe and chronic mental illness, however, have an emotional and practical impact on the caregiver. This is often referred to as the burden of care. The burden experienced by the family member affects not only their own mental, emotional and physical wellbeing, but also has a significant impact on the course of the patient's illness [3].

Some of the factors that contribute to increased burden are lack of awareness and knowledge about the illness, denial of the illness, self-blame, avoidance coping, including the use of tobacco and alcohol, and withdrawal from the social support network.

Disability Rehabilitation Management through ICT, 195–202.

The main goals in working with families is to achieve the best possible outcome for the patient through collaborative treatment and management and to alleviate the suffering of the family members by supporting them in their efforts to aid the recovery of the ill member [4].

Why is it Important to Help Families Cope?

There is overwhelming evidence that patients living with families do better in terms of their social and occupational functioning [5]. However, all families are not equally well equipped to support their family member with disability. There are sophisticated families with extensive resources at one end and families who are struggling with multiple problems and cannot offer any assistance to the ill member. The National Mental Health Programme seeks to deliver mental health care through a decentralized, community based primary health care network. This essentially means that families will continue to be the locus of care. Family members that use avoidance, denial and negative distraction techniques as coping strategies experience greater burden, while the use of problem solving, acceptance, religion and social support reduce the burden of care [6].

Most families do not have enough information about the patient's condition and how to deal with it. The lack of clarity around the process and outcome of illness results in many families regularly oscillating between hope and despair [7]. They feel helpless, vulnerable and alone. In the early stages, families often experience shame, embarrassment and self blame which may hinder help seeking and early intervention. In this phase, contact with a mental health professional is extremely important. As the illness persists, pessimism and despair replace the unquestioning faith in the doctor and the mental health system. At this point, other families who have been through the process are often perceived as more helpful than professionals. Family knowledge is an important resource in rehabilitation. Inherent strengths of family care are shared with other members of the community. When they are able to understand the limitations of what can be done, then they are more likely to focus on managing the symptoms of the illness and improve the functioning of the family member with the illness [8].

What Constitutes Family Intervention?

Family intervention programmes ideally comprise four components [9]:

(i) An educational component that provides information about the illness, the availability of services, the closest service provider, and so on.

(ii) A skills component that offers training in communication, problem solving, conflict resolution and behavioural management.

(iii) An emotional component that provides opportunities for sharing, grieving and mobilizing resources reduces the impact of the illness on family members through stress management.

(iv) A social component that strengthens social support coping and increases the utilization of informal and formal support networks.

The educational component includes the following:

- To teach caregivers about the symptoms and course of the mental illness.
- To provide family members the opportunity to ask questions about psychiatric disorders and treatment options. FAQs would find a place here.
- To help family members access mental health services at different levels of care by providing a resource directory and related information.
- To help manage medication and its side effects.
- To help family members disclose and talk about the illness with others in the family and social circle.

The skills component helps family members to cope with day-to-day problems such as:

- Hygiene and self-care
- Bizarre and abnormal behaviour
- Antisocial and aggressive behaviour
- Social withdrawal and isolation

- Self-destructive and suicidal behaviour
- Educational, vocational and career issues

The emotional component includes:

- Reduction of stigma of mental illness by providing a forum in which to discuss concerns and obtain support from peers.
- Taking care of oneself in order to be able to carry out the role of care provider.
- Address concerns of long-term care: Who or what after me?

One of the primary functions of peer support is to provide families with a place to share their stories about coping with mental illness. They can mourn their loss and feel validated in their experience. Even when there are no answers, families can learn to live better with what they cannot change.

The social component includes:

- Linking family members with opportunities for support both at the hospital and community centre, including local self-help groups or national associations.
- Understanding the rights and responsibilities of clients, family members and professionals.

Family support and self-help groups also facilitate the families playing a more active role by sharing their knowledge. Through advocacy, they feel empowered, confident and hopeful when they can influence policies and programmes that can positively impact on the lives of their family member with disability. Families want to be seen as partners in the rehabilitation process. Family-to-family education programmes have been found to increase the participant's knowledge about the causes and treatment of mental illness, their understanding of the health system and their own level of wellbeing [1].

The Role of ICT in Family Education and Coping

Information communication technology (ICT) is being increasingly used to extend the availability and range of mental health services. Some of the online services currently available include structured therapy programmes, psychological treatment by email, real-time online counseling, professionally-assisted chat rooms, self-help groups, health information and educational modules. A variety of computer technologies have been used including virtual reality, interactive voice response systems using telephony, the Internet and interactive television. In addition to the standard PC, palm-top computers, telephones, mobile telephones and television have been used. Computer-based treatment programs use state-of-the-art interactive, multi-media functionality, high-tech video graphics, animations and voiceover. The sessions are interactive and personalized. The user interface is specifically designed to be engaging and simple to use for people with no previous computer experience [11].

In India, rapid strides have been made in the use of ICT [12]. It has also been commented that India offers some unique characteristics in the use of social media, especially in the use of weblogs and mobile blogging [13]. This is especially relevant since mobile phone penetration has been far more than the personal computer.

ICT can be used to improve mental health literacy in caregivers. Information dissemination can be delivered through the Internet and user-friendly computer programs. Material should be easy to use, relevant and effective [14]. In India, with a large number of illiterate users, audio-visual components must be favoured, while in the west, most users are literate and access text material that can be downloaded and printed. Touch screen technology has been successfully used in presenting health information in a culturally relevant manner to indigenous communities in Australia [15].

Internet technology can be used to set up 'Meet the expert' hour during which the family members can directly interact with the treating professional or mental health consultant who can provide information and support specifically tailored to the needs of the family. Shanker [16] reported that families in India have little desire for theoretical information on the illness such as theories about etiology or the diagnostic terminology. What they look for is practical advice on handling day-to-day difficulties, symptom management and what they could do to improve the level of functioning of the patient. These are often best handled in an individualized programme. Direct interaction with the expert is also important as there is so much heterogeneity in the presentation of mental illness, waxing and waning course of symptoms, degree of disability that can vary from time to time, and individual patient's different responses to treatment and idiosyncratic responses to medication and its side effects.

Short educational video films can be screened at a community centre where a group meeting is conducted. This can also simultaneously be beamed to several centres through satellite. Family support groups can be initiated and sustained through such mechanisms. Computer-based mental health interventions have been found to be cost effective and clinically efficacious [17].

Who is the End User?

Family education is one of the steps in a stepped care approach. However, the end user may not always and necessarily be a family member, but a community health worker, nurse or trained community volunteer. ICT can be judiciously used in training of all the above care providers and in service delivery.

Conclusion

Health care policies the world over are acknowledging the need to make treatment accessible and equitable. While there can be no substitute for good clinical care and counseling, E-mental health care in India can significantly improve access to care by making it possible for families living even in remote areas to be in touch with other families as well as with professional help.

The gap between what we know and what we do for families with mental illness appears to be large [18]. Although initial costs of setting up ICT services may be high, it may help to overcome practical impediments such as transportation difficulties and competing time and energy demands on caregivers that emerge as barriers to accessing care. We must, therefore, continue to strive to develop practical and low cost strategies to provide family psychoeducation and enable families to cope with mental illness and disability. Access to information and resources can enable families of the mentally differently abled to get empowered.

References

[1] C. Barrowclough, 'Families of People with Schizophrenia', chapter 1 in N. Sartorious, J. Leff, J. J. Lopez-Ibor, M. Maj and A. Okasha (eds.), *Families and Mental Disorders: From Burden to Empowerment*, Chichester: Wiley & Sons, 2005.

[2] R. Shanker and K. Rao, 'From Burden to Empowerment: The Journey of Family Care-givers in India', chapter 12 in N. Sartorius, J. Leff, J. J. Lopez-Ibor, M. Maj and A. Okasha (eds.), *Families and Mental Disorders: From Burden to Empowerment*, Chichester: John Wiley & Sons. 2005.

[3] E. Kuipers and P. E. Bebbington, 'Research on Burden and Coping Strategies in Families of People with Mental Disorders: Problems and Perspectives', chapter 10 in N. Sartorius, J. Leff, J. J. Lopez-Ibor, M. Maj and A. Okasha (eds.), *Families and Mental Disorders: From Burden to Empowerment*, Chichester: John Wiley & Sons, 2005.

[4] I. Falloon, 'Research on Family Interventions for Mental Disorders: Problems and Per-spectives', chapter 11 in N. Sartorius, J. Leff, J. J. Lopez-Ibor, M. Maj and A. Okasha (eds.), *Families and Mental Disorders: From Burden to Empowerment*. Chichester: John Wiley & Sons, 2005.

[5] L. Dixon, 'Providing Services to Families of Persons with Schizophrenia: Present and Future', *Journal of Mental Health Policy and Economics*, vol. 2, 1998, pp. 3–8; Sherman, M.D., 'The Support and Family Education (SAFE) Program: Mental Health Facts for Families', *Psychiatric Services*, vol. 54, 2003. pp. 35–37.

[6] S. Chakrabarti and S. Gill, 'Coping and its Correlates among Caregivers of Patients with Bipolar Disorder: A Preliminary Study', *Bipolar Disorder*, vol. 4, 2002, pp. 50–60; L. Magliano, et al., 'Family Burden and Coping Strategies in Schizophrenia: 1 Year Fol-lowup Data from the BIOMED I Study', *Social Psychiatry & Psychiatric Epidemiology*, vol. 35, 2000, pp. 109–115; A. Rammohan, K. Rao, and D. K. Subbakrishna, 'Religious Coping and Psychological Wellbeing in Carers of Relatives with Schizophrenia', *Acta Psychiatrica Scandinavica*, vol. 105, 2002, pp. 356–362.

[7] L. Spaniol, A. M. Zipple, and D. Lockwood, 'The Role of the Family in Psychiatric Rehabilitation, *Schizophrenia Bulletin*, vol. 18, 1992, pp. 341–348.

[8] Ibid.

[9] D. T. Marsh, *Families and Mental Illness: New Directions in Professional Practice*, New York: Praeger, 1992.

[10] L. Dixon, 'Providing Services to Families of Persons with Schizophrenia: Present and Future', *Journal of Mental Health Policy and Economics*, vol. 2, 1998, pp. 3–8; L. Dixon, et al., 'Evidence Based Practices for Services to Families of People with Psychiatric Disabilities, *Psychiatric Services*, vol. 52, 2001, pp. 903–910.

[11] K. M. Griffiths and H. Christensen, 'Internet-based Mental Health Programs: A Power-ful Tool in the Rural Medical Kit', *Australian Journal of Rural Health*, vol. 15, 2007, pp. 81–87; J. Proudfoot, 'Computer-based Treatment for Anxiety and Depression: Is it Feasible? Is it Effective?' *Neuroscience and Biobehavioural Reviews*, vol. 28, 2004, pp. 353–363.

[12] S. C. Bhatnagar and R. Schware, *Information and Communication Technology in Devel-opment Cases from India*, New Delhi: Sage Publications, 2000; A. Singhal, and E.M. Rogers, *India's Information Revolution: From Bullock Carts to Cyber Marts*, New Delhi: Sage Publications, 2001.

[13] Han Kyung, 'Indian Social Media: A Lesson for the West', *Deccan Herald*, 5 February, p. 8.

[14] M. D. Sherman, 'The Support and Family Education (SAFE) Program: Mental Health Facts for Families', *Psychiatric Services*, vol. 54, 2003. pp. 35-37.

[15] D. P. Doessel, H. Travers, and E. Hunter, 'The Use of Touch-screen Technology for Health-related Information in Indigenous Communities: Some Economic Issues', *Prometheus*, vol. 5, 2007, pp. 373–392.

[16] R. Shanker, 'Interventions with Families of People with Schizophrenia in India', *New Directions in Mental Health Services*, vol. 62, 1994, pp. 79–88.

[17] I. Marks, 'Computer Aids to Mental Health Care', *Canadian Journal of Psychiatry*, vol. 44, 1999, pp. 548–555.

[18] E. Susser, P. Collins, B. Schanzer, V. K. Varma, and M. Gittelman, 'Can We Learn from the Care of Persons with Mental Illness in Developing Countries?' *American Journal of Public Health*, vol. 86, 1996, pp. 926–928.

20

ICT for Educational Audiology and Education of Persons with Hearing Impairments

R. Rangasayee[1], Varsha Gathoo, Rajeev Jalvi[2] and P.M. Mathew*

[1] *Director, Ali Yavar Jung National Institute for the Hearing Handicapped, Mumbai*
[2] *Head, Department of Audiology, Ali Yavar Jung National Institute For The Hearing Handicapped, K.C Marg, Bandra Reclamation, Bandra (West), Mumbai-400050, India*

Background

Persons with hearing impairment can perform any task like their peers who can hear normally. However, as there is discrepancy in the performance of a task within hearing persons, such variations can and do exist among persons with hearing impairment. Certain enabling provisions can facilitate equal or greater participation of either neglected or disadvantaged persons.

The Internet with its powerful penetration and scalability has become an increasingly popular medical information resource and a new platform for tele-health. It is possible to create Internet-based tele-audiological systems which will enable audiologists to perform real-time assessment of pure-tone auditory thresholds and otoacoustic emissions through remote control.

A key enabling provision with potential for universal design is ICT. It is in this context that this article, besides exploring possible solutions, also addresses issues pertaining to hearing impairment through ICT.

*Director; Varsha Gathoo is Reader/HOD, Education; Rajeev Jalvi is Reader/HOD, Audiology; and P. M. Mathew is Computer Programmer, at AYJNIHH, Mumbai

Disability Rehabilitation Management through ICT, 203–214.

Bridging the Digital Divide

(i) The number of persons by general education per 1,000 disabled persons in the age group 15 years and above is given in Table 1 (NSSO report, 2002).

Type of Disability	General Education					
	Not Literate	Primary	Middle	Secondary	Higher Secondary	Graduation and Above
Locomotor Disability	465	226	154	77	47	31
Blindness	788	122	45	25	11	09
Hearing Disability	688	195	76	36	16	09
Mental Retardation	862	95	37	03	02	01
Any disability	496	243	145	63	35	16

PWD form a very heterogeneous group. Each disability is unique in its characteristics and hence experiences differential barriers within its group. Perhaps acquisition of literacy skills possesses the greatest challenge for people with hearing impairment (PWHI), because of their associated problems of language, speech and communication, leading to difficulty in reading comprehension and written expression. Hence, school and college drop-out rate among them increases.

(ii) ICT, like education, is a leveler. While ensuring the use of technology, one needs to take care that it does not become a source of further inequality.

The digital divide should not accentuate the already existing disparities and further sideline marginalized groups of learners.

(iii) The Millennium Development Goal (2000) emphasizes the eradication of poverty and recognizes the potential of technology for development [6]. The UNESCAP (2002) also adopted the resolution of promoting inclusive and barrier-free environment for PWD. The Biwako Millennium Framework has listed access to ICT as an important strategy towards an inclusive and barrier-free society. The UNESCO's ICT in Education policy promotes appropriate policy models and strategies for proper integration of ICT into the teaching-learning process.

(iv) The digital divide amongst PWHI further deepens with respect to gender, socio-economic factors and demography. A study conducted by Kharbyngar (2004) indicates students with hearing impairment significantly lag behind their hearing counterparts in word processing (computer application) at knowledge, understanding and skill levels. However, in the interest of country's social and economic growth, it is but essential to bridge this divide.

Rehabilitation of PWHI: Issues and E-measures

The key components for the participation and equalization of opportunities are;

- Physical restoration
- Informed learning
- Social networking

These issues in turn lead to personal, social and economic growth and development of an individual. ICT has the potential to address these issues through a 'cafeteria approach'.

Physical Restoration

Manifestation of hearing loss at any given stage of life affects three 'e's, viz., education, entrepreneurship/employment and entertainment. In order to reduce the detrimental effects of hearing loss, prevention, early identification and intervention are the core issues. Access to information about services for screening, diagnosis, certification, fitment and maintenance of aids and appliances will assist in physical restoration of hearing loss.

E-health

E-health is a relatively new area of interest that pertains to all health professions. It is described as 'an emerging field in the intersection of medical informatics, public health and business, referring to health services or information delivered or enhanced through the Internet and related technologies. In a broader sense, the term characterizes not only a technical development, but also a state of mind, a way of thinking, an attitude, and a commitment for networked, global thinking, to improve health care locally, regionally and worldwide by using information and communication technology' (Eysenbach, 2001).

Tele-health, tele-practices, computer-based health communication and interactive health communication all are terms commonly used to describe the use of information and communication technologies to provide health services from a distance. E-health also encompasses the use of technologies for clinician-to-clinician and for client-to-client communications.

E-mail has been used as a means of communication between clinicians and clients for many years (e.g., Johston, 1996). Hobbs et al. (2003) showed that 68 per cent of the 71 general physicians they surveyed used e-mail with some of their clients. It was also suggested by the American Medical Informatics Association that e-mail followup allows retention and clarification of advice provided in clinic.

Audiology and E-health

Audiology clients also use the Internet. In 2001, 34 per cent of American hearing-aid owners had access to email (Kochkin, 2002). More recently, 64 per cent of the clients of a London ear, nose and throat clinic reported having access to the Internet (Tassone et al., 2004). In 2002, 12 per cent of audiologists surveyed by the American Speech and Hearing Association reported using tele-practice (American Speech and Language Hearing Association, 2002), with the term tele-practice being defined as 'the application of technology to deliver audiology and speech language pathology services at a distance' (American Speech and Language Hearing Association 2002, p. 1). The vast majority of audiologists using tele-practice also reported that they usually used it to contact their clients while the clients were at home. The most frequent types of services provided were counseling (83 per cent) and follow up (68 per cent). The Internet was also used to deliver cognitive behavioural therapy to participants affected by tinnitus.

With the rapid development of Internet-based applications, tele-health has the potential to provide important healthcare coverage for rural areas where specialized audiological services are lacking. Various Internet applications such as intra-operative monitoring, fitting and programming of digital hearing aids, activation and mapping of cochlear implants, audiological diagnostic testing with otoacoustic emissions and middle ear immittance measures, and auditory training are but a few examples of potential areas where tele-hearing can be implemented through the Internet.

The advent of modern communication technology has unleashed a new wave of opportunities and threats to the delivery of healthcare services. Tele-health is the use of information and communication technology to transfer information and/or data in support of healthcare. The American Speech-Language Hearing Association (2001) defines tele-health as 'the application of technology to deliver health services at a distance by linking clinician and patient or clinician and clinician to provide any or all of the following:

1. Training, counseling, education
2. Assessment–establishing patient status
3. Intervention treatment/management

and to provide remote support and training of practitioners.' This area is in the early stages of development but is beyond the conceptual and start-up phase. Tele-health has impacted nearly all aspects of healthcare and has reached around the world. Health professionals can communicate quickly, widely, and directly with patients and colleagues, no matter where they are. It is only in the last few years that investigations in tele-health have moved from working with prototypes that test clinical possibilities to evaluating the utility and validity of new systems in the clinical setting. The impact of tele-health in healthcare could be tremendous, including the possibility of providing quality healthcare through telecommunications technology to underserved populations in prisons, inner cities, and rural locations and to elderly people in old age homes. Tele-health practitioners now have the potential to make a global impact on healthcare delivery in their respective professions.

The Internet and Telehealth

In recent years, the Internet, with its powerful penetration and scalability, has become an increasingly popular medical information resource and a new platform for tele-health. It took the radio 38 years to reach 50 million users and television 13 years, but the Internet reached similar numbers in only 5 years.

The Internet has already been successfully used in varied medical applications such as radiology, cardiology, psychiatry, social work, and dentistry.

So it seems intuitive to use the Internet as a vehicle to reach and provide audiological services to consumers. The ASHA recognized the potential of the Internet in providing services and emphasized the need for ASHA and its members 'to improve their use of Web-based and advanced technology to enhance the provision of personnel preparation, provision of clinical services including tele-practices and program administration'. Audiologists have been using some kind of tele-health practices on a regular basis in a limited manner for diagnostics, hearing aid fittings and counseling.

Audiological clinics employ websites that provide information about various services offered and provide contact information. E-mails facilitate scheduling appointments and peer consultations and tele-conferences allow for audiological clinical discussion boards.

Telehearing Systems

Tele-audiometry system: It is possible to develop a commercially available audiometer via an Internet protocol (IP) network which will allow an audiologist to conduct a hearing test remotely using a desktop or palmtop computer and an Internet connection.

Briefly, the tele-audiometry system can comprise three primary subsystems:

1. A controller that interfaces at a low level with an audiometer
2. A micro webserver to manage the IP connection and act as a traffic controller between the microcontroller and the remote user/consultant (i.e. the audiologist)
3. A remote computer, either a desktop personal computer (PC) or palmtop personal digital assistant (PDA) that uses a specially developed Microsoft Windows or Palm OS client program.

Figure 20.1 illustrates the main components of this tele-audiometric system.

The micro-web server and the controller will allow the audiometer's settings to be manipulated through a remote computer. The web server's functionality will enable it to send and receive information between the controller and the audiometer simultaneously, while hosting a web server for connection with the client PC or PDA. The micro-web server will command the controller while verifying that change is made using the output from the audiometer. This will

209

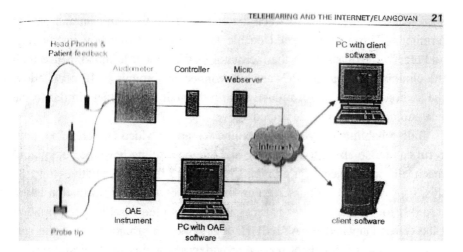

Fig. 20.1 Key components of the tele-audiometric system.

provide redundant verification that the audiometer is using the correct intensity and frequency settings, as well as which ear is receiving audio tones. This is essential to verify that the subject is actually receiving the correct tones. The following can be controlled with the Windows program: ear routing, air/bone conduction, transducer (insert or head phone), stimulus interrupter, stimulus lock, frequency modulation, pulsed stimulus, and masking. The micro-web server will allow an audiologist to control the audiometer from any location within an IP connection. Therefore, the system could be operated over the Internet, within an isolated local or wide area network within a health system, or via a dial up connection. The system can be designed in such a way so that it can be applied to any audiometer.

Tele-OAE System: The tele-OAE system can be designed using two PCs, one at the clinician terminal and the other at the patient terminal. A remote control software (PC Duo) can be installed in both PCs. The PC Duo software will allow remote access of the PC at the patient terminal through the PC at the clinician terminal with both PCs connected through an IP connection. The PC at the patient terminal will be installed with a PC-based OAE system which will be capable of performing diagnostic DPOAE testing. Through the PC Duo software, the clinician at the remote terminal will have complete access to the PC at the patient terminal and thus could initiate DPOAE testing and view the test results. The PC Duo software has provisions to send real-time messages between the computers at the clinician and patient terminals. This facility would enable the clinician to convey instructions and receive feedback from the facilitator at the patient terminal. Since both the clinician and patient PCs are interconnected through an IP connection, the audiologist can perform OAE testing through the Internet similar to the tele-audiometric system described above.

Tele-rehabilitation in terms of online screening, video conferencing, tele-certification, so also tele-mentoring and tele-counseling have the potential to reach 'all'. www.checkhearing.nic.in, is one of AYJNIHH's initiatives towards this. It provides a hearing screening through the website. Information about referrals for further services is accessible through the Disability Information Line (DiL) launched by AYJNIHH. It provides information about the service schemes and details of service providers pertaining to all disabilities, in the areas of need and interest. This operates on an IVRS currently covering Mahrashtra, Goa (tel. no. 022-264040019, 26404024, 26404043, 26405454) and New Delhi (tel. no. 011-29825094/95/96).

Informed Learning: Providing and sustaining quality education is an essential component of a knowledge-based society. It involves solving the triangle of quality, quantity and equity. The shift in education also has to be from 'teaching' to 'learning' and from 'rote memorization of content' to 'creativity and entrepreneurship'. Since parents and teachers are viewed as support partners, mobilizing them into lifelong learning groups is essential in the knowledge-based society. Technology mediation in education is considered essential to offer quality education at reduced cost with access to any one, any where and any time. ICT offers not only convergence of technologies, but also convergence of existing educational systems of formal, non-formal and informal education [11].

ICT in the form of computer-assisted instructions, interactive learning packages, subtitling of audio-video learning materials have a learner-centric approach and are especially benefiting learners with hearing impairment. AYJNIHH has developed prototype of adapted textbooks in the form of CDs.

A study conducted by Gathoo, Maricar and Mathew (2003) indicates significant differences in the attainment of students with hearing impairment at knowledge, understanding and application levels with computer-assisted instructions [1].

CART and C-print technologies are reported to supplement the teaching-learning process. Open distance learning (ODL) can address the issues in accessibility of education. To facilitate distant learning, virtual open schools, e-conferencing, so also Direct To Home (DTH) technology can connect schools to home. Software to support learners with speech and hearing difficulties like digitized speech facility, talking word processors, talking books enabling pupils who are unresponsive and avoid conversations, word processors with banks of words and phrases, overlay keyboards, vocabulary spelling and grammar programs, literacy assistance software, and so on could be used.

Social Networking: For an informed society, individuals need to be equipped with knowledge about the social structure, relations and interactions. They also need to be aptly informed about the current affairs, global trends, scope of employment and emerging avenues of job market. Further, information about the facilities, provisions and concessions would prevent exploitation and marginalization. Networking would bring about a revolution in information transformation.

Accessible ICT in the form of wireless handhelds like mobile phones, instant messenger, real-time captioning, e-mail, speech-to-text technology are useful for communication between persons with hearing impairment. Video phone booths are useful for the video relay service using sign language interpreters. Accessibility services like LAMA (Location Aware Messaging for Accessibility) need to be launched at public places for dissemination of information. Knowledge portals could be created for pooling and accessing information [10]. Social networking websites would help persons with disabilities to contribute online. A website jobsfordeaf.nic.in launched by AYJNIHH facilitates both employers and persons with hearing impairment to gain employment. The creation of virtual communities would provide opportunities for face-to-face interaction and to increase participation. Accessible knowledge parks and infotech parks could be redesigned as centres for empowerment for persons with hearing impairment.

Initiatives

We suffer more often in apprehension than reality. The Hole-in-the-Wall experiment by Dr. Mitra of NIIT proves that literacy or socio-economic factors are no barriers to ICT. In the experiment, a high-speed computer with Internet connectivity was placed in the wall separating the NIIT office and a slum. The web camera documented the speed with which the slum children started browsing and eliciting information. This encouraging experiment highlights the fact that lack of proficiency in English or literacy per se does not lead to digital divide.

The following recommendations are suggested to address the issues discussed:

- Creating universal designs for all
- Technology that makes access inexpensive [9]
- Content to be more graphic
- Induction of ICT in curriculum and making it mandatory
- Capacity building in terms of ICT training to teachers and service providers [5]
- Development of performance indicators to monitor outcomes of application of ICT
- C-print technology for higher education
- Software for assessment of hearing and hearing technologies
- Speech-to-text technology addressing multilingual issues
- Research in etechnology applications to increase the reach with reference to diagnostics testing as well as tele-rehabilitation/therapy

Conclusion

These systems can remove the distance barrier and allow for remote assessment of hearing and provision of audiological care without an audiologist on site. Although these investigations demonstrate the feasibility of using tele-practice models to provide clinical audiological services, there are several issues that one needs to address prior to using these systems to deliver services at a distance. These issues include training of personnel to function as onsite facilitators in the remote test sites, ensuring compliance with laws and regulations of relevant jurisdictions regarding professional service delivery (e.g., the Health Insurance Portability and Accountability Act) and licensing, candidacy criteria, privacy and confidentiality of the patient and test results and other potential technical issues related to operating under optimal conditions. Though clinically trained audiologists perform the role of the facilitator at the remote test site, however, in an ideal tele-practice setup, the clinician will not be present in the test site while an appropriately trained individual (e.g., paraprofessional, health professional, trained school teacher) will serve as the facilitator. The facilitator's role includes, but is not limited to, assessing candidacy criteria, building rapport with the patient, taking initial case history, providing information about test procedures, placing headphones or probe tips, providing feedback and coordinating with the clinician at the remote test site.

Another concern is the transmission of patient information and test results over distances through some electronic medium such as the Internet. Practitioners of tele-practices also share ethical and legal responsibility of protecting and preserving the privacy and confidentially of their patients. A caution must be exercised when patients are tested from a distant location and information is transmitted through a wider and potentially insecure network (e.g., the Web). Electronic safety measures such as data encryption, firewall, website certification, advanced authentication of users and use of secure private networks can be incorporated to ensure compliance with HIPAA regulations.

The present climate of clinical audiology demands that we must enhance the quality of service while controlling the cost of its delivery. Tele-health practitioners in the field of communication sciences and disorders have essentially used the so-called information highway for purposes such as disseminating information, scheduling and maintaining patient and consultant contacts, and efficient transfer of health records. Although global exchange of information is considered one of the major sources of scientific progress in healthcare, the Internet has more potential than what has been tapped so far. With computers and the Internet being nearly ubiquitous, Internet-based applications in practically all kinds of health setups are feasible. Indeed, the ultimate success of tele-health in the field of audiology depends on how well the hearing care providers exploit the capabilities of advanced information technology and find ways to adapt, transform and implement them into our present hearing healthcare system.

References

[1] American Speech Language Hearing Association, 'Tele-practices and ASHA: Report of the Tele-practices Team 2001', available at http://professional.asha.org. Accessed 27 March, 2003.
[2] American National Standards Institute, 'Maximum Permissible Ambient Noise Levels for Audiometric Test Rooms', ANSI S3.1-1991, New York: ANSI, 1991.

[3]   American Speech Language Hearing Association, 'Guidelines for Identification Audiometry', ASHA, vol. 27, 1985, pp. 49–52.

[4]   R. O. Bashshur, 'Telemedicine and the Health Care System', in R. Bashshur, J. H. Sanders, G. W. Shannon (eds.), *Telemedicine: Theory and Practice*, Springfield, IL: Charles C. Thomas, 1997, pp. 5–33.

[5]   R. L. Bashshur, 'Telemedicine/Telehealth: An International Perspective', *Telemedicine and Health Care,* Telemed J E Health 2002, 8, pp. 5–12.

[6]   EDF contribution to the 3rd Preparatory Committee Meeting, World Summit on the Information Society, http://www.edffeph.org.

[7]   S. Elangovan, 'Telehearing and the Internet', *Seminars in Hearing*, vol. 26, no. 1, 2005.

[8]   Gathoo, et al., 'Application of ICT in Classroom Learning for Children with Hearing Impairment', The 23rd Asia Pacific International Seminar on Special Education, Japan, 2003.

[9]   D. G. Givens and S. Elangovan, 'Internet Application to Tele Audiology "Nothin" but Net', American Journal of Audiology, 12, 2003, pp. 1–8.

[10]   'How can ICT Play a Role?' http://www.i4donline.net

[11]   'Hole-In-The-Wall Training Systems', http://hole-in-the-wall.com/Beginnings.html

[12]   'ICT Services for Disadvantaged', http://www.dirf.org/ictd.htm

[13]   'I-CONSENT Virtual School and Learning Homes (VSLH)', Brochure of I-CONSENT programs and pilot of VSLH with Tech-MODE Approach.

[14]   Internet   Society   Disability   and   Special   Needs   Chapter, http://www.worldenable.net/manila203/overview.htm

[15]   *Journal of Technology Management and Innovation* http://www.jotmi.org

[16]   Kharbyngar, 'Attainment of Students With and Without Hearing Impairment in Word Processing', unpublished.

[17]   M. Krumn, J. Ribera, T. Froelich, 'Bridging the Service Gap through Audiology Telepractice' *ASHA*, 7, 2002, pp. 6–7.

[18]   D. McCrty and C. Clancy, 'Tele-health: Implications for Social Work Practice', *Social Work*, 47, 2002, pp. 153–161.

[19]   Ministry of Statistics and Program Implementation, Government of India, Millennium Development Goals in India, report 2005 http://www.undp.org.in

[20]   *National Sample Survey Organization, Ministry of Statistics and Programme Implementation*, Government of India (2002), Disabled persons in India

[21]   'Policies on Disabled Persons and ICT', Backgrounder en:wikipedia.org/wiki/web_Accessibility_Initiatives_in_phillipines_

[22]   J. M. Birnbach, 'The Future of Tele-dentistry', J. Calif Dent Assoc, 28, 2000, pp. 41–143.

Index

Lightning Source UK Ltd.
Milton Keynes UK
UKOW02n0843151214

243137UK00001B/43/P